Handbook of Human Factors in Cybersecurity Systems

Data and security breaches can cost organizations millions and result in bad publicity and lost income, with humans often being regarded as the "weakest link" in cybersecurity incidents. This book is a cross-disciplinary primer on the latest research on human aspects of cybersecurity, reimagining the role of humans in cybersecurity systems. It aims to increase participation in research on the emerging, critical, and impactful area of human factors of cybersecurity by orienting human factors researchers, professionals, and students to cybersecurity issues while providing them with comprehensive, cross-cutting knowledge from human factors that applies to this emerging area.

Across nine chapters, the *Handbook of Human Factors in Cybersecurity Systems* covers the essential topics for understanding humans to optimize cybersecurity system performance. Topics included cover system safety, cybersecurity teams, safety culture, cyberattacks, defense approaches, phishing, emerging cybersecurity threats, and many more. The book investigates cybersecurity across numerous industries, including aviation and healthcare, and is backed up by interesting, real-life case studies. The reader will develop a full and thorough understanding of how to integrate human factors principles with cybersecurity practices.

With chapters from a team of global experts, this book is an essential read for any student, academic, or professional in the disciplines of computer engineering, systems engineering, safety engineering, human factors, and ergonomics and cybersecurity engineering.

The Human Element in Smart and Intelligent Systems

Series editor: Abbas Moallem

This series aims to cover the role of the human element in all aspects of smart and intelligent systems. It will include a broad range of reference works, textbooks, and handbooks. The series will look for single-authored works and edited collections that include, but not limited to: automated driving, smart networks and devices, cybersecurity, data visualization and analysis, social networking, smart cities, smart manufacturing, trust and privacy, artificial intelligence, cognitive intelligence, pattern recognition, computational intelligence, and robotics. Both introductory and advanced material for students and professionals will be included.

Smart and Intelligent Systems: The Human Elements in Artificial Intelligence, Robotics, and Cybersecurity
Abbas Moallem

Understanding Cybersecurity Technologies: A Guide to Select Right Cybersecurity Tools
Abbas Moallem

Human Factors Analysis of 23 Cyberattacks
Abbas Moallem

Handbook of Human Factors in Cybersecurity Systems: A Human-Centered Approach
Edited by Crystal M. Fausett, Joseph R. Keebler, David Schuster and Elizabeth H. Lazzara

For more information on this series, please visit: https://www.routledge.com/The-Human-Element-in-Smart-and-Intelligent-Systems/book-series/CRCHESIS

Handbook of Human Factors in Cybersecurity Systems

A Human-Centered Approach

Edited by

Crystal M. Fausett, Joseph R. Keebler,
David Schuster, and Elizabeth H. Lazzara

CRC Press
Taylor & Francis Group
Boca Raton London New York

CRC Press is an imprint of the
Taylor & Francis Group, an **informa** business

Front cover image: Vit-Mar/Shutterstock

First edition published 2026
by CRC Press
2385 NW Executive Center Drive, Suite 320, Boca Raton FL 33431

and by CRC Press
4 Park Square, Milton Park, Abingdon, Oxon, OX14 4RN

CRC Press is an imprint of Taylor & Francis Group, LLC

ISBN: 978-1-032-52493-1 (hbk)
ISBN: 978-1-032-53804-4 (pbk)
ISBN: 978-1-003-41370-7 (ebk)

DOI: 10.1201/9781003413707

Typeset in Times LT Std
by KnowledgeWorks Global Ltd.

Contents

About the Editors

Crystal M. Fausett is Assistant Professor in the School of Information at San José State University, USA, specializing in human-technology interaction and systems design. With expertise spanning healthcare, cybersecurity, and military operations, her research examines how diverse user groups interact with complex systems and process critical information under varying conditions. As an advocate for human-centered approaches to technology, she is dedicated to developing systems that accommodate human capabilities while mitigating limitations, transforming users from "the weakest link" into essential partners. Dr. Fausett holds a PhD in Human Factors from Embry-Riddle Aeronautical University, USA.

Joseph R. Keebler is Professor of Human Factors at Embry-Riddle Aeronautical University, USA, and has over 20 years of experience conducting experimental and applied Human Factors science, with a specific focus on training and teamwork in medical, military, and consumer domains. He has partnered with multiple agencies leading projects aimed at the implementation of HF/E in complex, high-risk systems to increase safety and human performance. He directs the Small Teams Analog Research (STAR) Laboratory, and co-directs the Research Engineering and Applied Collaborations in Healthcare (REACH) Laboratory. These laboratories include a team of faculty and students who work to solve real-world teamwork and training issues, with STAR focusing on training and teamwork in aerospace, cybersecurity, robotics, and artificial intelligence, and REACH focusing on patient safety, training, and teamwork issues in healthcare settings. He holds a PhD in Psychology from the University of Central Florida, USA.

David Schuster is Professor in the Department of Psychology at San José State University, USA, and Principal Investigator at the Virtual Environments, Cognition, and Training Research (VECTR) Lab. Dr. Schuster's research centers on understanding individual and shared cognition in complex environments. He has conducted research in domains such as aviation, transportation security training, and human-robot interaction. He is interested in how complex sociotechnical systems support or hinder people, with a focus on decision-making among cybersecurity professionals. He holds a PhD in Psychology from the University of Central Florida, USA, specializing in Applied Experimental and Human Factors Psychology.

Elizabeth H. Lazzara is Associate Professor at Embry-Riddle Aeronautical University, USA. She currently co-leads the Research, Engineering, and Applied Collaboration in Healthcare (REACH). Although she has extensive experiences in military, academic, and commercial settings, her primary interests lie within improving the quality of patient care within the health-care domain. Dr. Lazzara strives to make a long-lasting and significant impact by examining and advancing the science and practice of clinical care and patient safety issues pertaining to human performance, teamwork, team training, simulation-based training, and performance measurement. She holds a PhD in Human Factors Psychology from the University of Central Florida.

Contributors

Palvi Aggarwal
0000-0003-2488-8959
USA

Lila Berger
0009-0002-8852-5129
USA

Leonie Boskeljon-Horst
0000-0001-5760-6191
The Netherlands

Liam D. Brennan
0009-0008-1333-9421
USA

Gabriela Fernández Castillo
0009-0001-9317-1858
USA

Benjamin A. Clegg
0000-0001-6026-5076
USA

Jonathan Dekker
0009-0004-2515-2286
The Netherlands

Sidney Dekker
0000-0002-9141-3374
Australia

Crystal M. Fausett
0000-0002-1458-9763
USA

Sarah D. Fouquet
0000-0002-9144-5165
USA

Robert S. Gutzwiller
0000-0002-2159-8131
USA

Giovanna Henery
USA

Xiaojian Jin
0009-0008-1468-0381
USA

Joseph R. Keebler
0000-0003-2246-7472
USA

Elizabeth H. Lazzara
0000-0002-3495-0595
USA

Christina M. Lewis
0000-0003-0868-4999
USA

Rylee Linhardt
0009-0009-6779-016X
USA

Yongxin Liu
0000-0003-4527-8623
USA

Hannah C. Luckes
0009-0000-8418-5775
USA

Rosa Martey
0000-0003-4818-6820
USA

Rebecca Pharmer
0009-0007-9221-1698
USA

Indrakshi Ray
0000-0002-0714-7676
USA

Eduardo Salas
0000-0002-5448-4958
USA

David Schuster
0000-0002-2698-7654
USA

Richard J. Simonson
0000-0003-4272-1570
USA

Houbing Song
0000-0003-2631-9223
USA

Jian Wang
0000-0001-6043-6971
USA

Stephen J. Zaccaro
0000-0002-8313-2206
USA

Joanna Zakzewski
0000-0003-3497-8675
USA

Preface

At the time, I had no idea that my presentation at the Human Factors and Ergonomics Society Annual Meeting would transform into the book you now hold into your hands. What began with the simple intersection of interests related to cybersecurity as well as handoff processes would blossom into how human factors principles can revolutionize our approach to cybersecurity.

The traditional narrative surrounding humans in cybersecurity has positioned them as vulnerable, often labeling them as the "weakest link." This book, I believe, challenges that narrative. We propose a fundamental shift: from viewing humans as weaknesses to recognizing them as essential partners in creating robust cybersecurity environments.

This book is being published at a critical time. As cyber threats grow increasingly sophisticated, the limitations of purely technical approaches become more apparent. By integrating human factors principles with cybersecurity practices, we can develop more effective strategies that acknowledge both limitations and capabilities of humans. The chapters that follow explore this integration through various lenses, from sociotechnical systems perspectives to team effectiveness models, from organizational resilience to innovative defense approaches that leverage human psychology. Our authors comprise a diverse network of industrial organizational psychologists, safety professionals, human factors scientists, and experts working at the intersection of cybersecurity and domains like aviation and transportation. This multidisciplinary perspective is one that we believe to be essential for addressing the complex cybersecurity challenges that we face today.

Human factors professionals seeking insight into cybersecurity and cybersecurity researchers and practitioners alike will find actionable strategies within these pages. The field of human factors in cybersecurity is growing and evolving, and this textbook represents a step in what promises to be an exciting area of exploration.

It is a great honor to publish this alongside my coauthors. Their research ethos has profoundly shaped my own approach to human factors in cybersecurity. Our collective experiences have been instrumental for this textbook. It is my hope that this textbook contributes to a more nuanced understanding of the human dimension in cybersecurity and inspires innovative approaches that recognize humans not as liabilities but as essential partners in our cybersecurity efforts.

Dr. Crystal M. Fausett

You Can't Take the Human Out of Cybersecurity

Reimagining People as Partners, Not Problems

Crystal M. Fausett, David Schuster, Elizabeth H. Lazzara, and Joseph R. Keebler

INTRODUCTION

Traditional approaches to cybersecurity have largely focused on technical solutions; while, relegating humans to the status of "the weakest link" in the cybersecurity chain (Goh, 2021; Yan et al., 2018). This perspective, while not entirely without merit, represents an incomplete and counterproductive understanding of the cybersecurity landscape. This textbook challenges this conventional wisdom by reimagining the role of humans in cybersecurity systems. Rather than viewing humans merely as vulnerabilities to be managed or controlled, we propose a shift in thinking toward understanding humans as essential partners in creating strong cybersecurity environments. Robust cybersecurity outcomes depend not only on technological sophistication but also on the complex interactions between people, technology, processes, and organizational structures.

This is not to say that humans don't have limitations in the cybersecurity context; we undoubtedly do. Our limited capacity for memory makes managing the average person's 90+ unique passwords virtually impossible, and our susceptibility to social engineering tactics creates exploitable vulnerabilities (Pujari & Hussain, 2024; Rooparaghunath et al., 2020). However, our capabilities are significant and often undervalued. Unlike technological systems that simply execute their programming, humans excel at pattern recognition, detecting anomalies, and making intuitive judgments about when something feels "wrong" (Jain & Duin, 2004). We possess the ability to intervene when technology goes awry, interpret ambiguous situations, and understand the broader implications of cybersecurity breaches or practices in ways that technological systems cannot replicate.

This inherent tension between human cybersecurity limitations and capabilities points to a natural conclusion: effective cybersecurity requires leveraging human capabilities while simultaneously accommodating our limitations. Addressing this challenge requires collaboration across all sectors of society and a reliance on the science that studies human-technology interactions, namely the field of human factors. Human factors is an interdisciplinary field that blends elements of psychology

DOI: 10.1201/9781003413707-1

and engineering while applying scientific knowledge about human limitations and capabilities to the analysis and design of systems to ensure that total system performance is optimized (Stone et al., 2017).

For instance, in aviation, human factors research has contributed to the development of flight deck designs that enhance pilot situational awareness, reduce errors, and improve safety through features such as automated alerts and ergonomic controls (Keebler et al., 2022). Similarly, in healthcare, human factors principles have been applied to the design of electronic health records and medical devices to minimize errors and improve patient outcomes. These successes demonstrate the transformative potential of human factors in optimizing complex systems (Cafazzo & St-Cyr, 2012). The human factors approach considers human cognition, perception, attention, decision-making, and motor control and integrates them into system design for the purposes of enhancing reliability, efficiency, safety, and usability (Stone et al., 2017). Human factors creates systems that align with humans with the goal of reducing errors and improving outcomes. With a proven track record in fields like aviation and healthcare, human factors is uniquely suited to address the complex challenges that face cybersecurity today.

As cyberthreats continue to evolve, our approach to cybersecurity must evolve as well. By integrating human factors principles with cybersecurity practices, organizations can begin to dismiss overly-simplistic blame narratives in favor of nuanced and effective cybersecurity strategies. This textbook offers both theoretical frameworks and practical guidance for this transformation, serving as a handbook for cybersecurity professionals, human factors scientists, and organizational leaders committed to building resilient cybersecurity systems in the digital frontier.

KEY THEMES FROM THIS TEXTBOOK

From "Human as Weakness" to "Human as Partner"

Throughout this book, you will be presented with calls for a reconceptualization of how we view the human role in cybersecurity. Traditional approaches have often painted humans as "the weakest link" in cybersecurity systems. Humans are seen as vulnerabilities. They are prone to errors, susceptible to social engineering, bad at creating safe passwords, and bypass security measures that are too cumbersome. While there is no denying that these vulnerabilities exist, our contributors demonstrate that viewing humans only as liabilities is both inaccurate and counterproductive. As Schuster argues in the opening chapter, we must shift our thinking toward a partnership model, in which humans are seen as essential collaborators in cybersecurity efforts. Similarly, Boskeljon-Horst, Dekker, and Dekker challenge us to view human error not as a *cause* of security failures but as a *symptom* of deeper systemic issues. This perspective encourages cybersecurity professionals to look beyond individual blame toward organizational, technological, and procedural factors that shape human behavior. This shift in thinking is not just a semantic change. It would fundamentally alter the way in which we view cybersecurity incidents, train personnel, and cultivate security culture.

Sociotechnical Systems Perspective

At the heart of this book is the application of sociotechnical systems theory to cybersecurity challenges. This approach highlights that cybersecurity failures rarely result from isolated technical or human deficiency but rather manifest from the complex interactions that occur between people, technology, process, and organizational structure. By connecting cybersecurity to similarly high-risk domains such as aviation and healthcare, Schuster proposes the adoption of established sociotechnical frameworks. These frameworks serve as a lens through which cybersecurity professionals and other stakeholders may consider broader organizational and contextual factors that influence cybersecurity behaviors and outcomes. Rather than emphasizing technical controls or individual compliance, this perspective promotes comprehensive interventions that address system components.

Team Effectiveness and Multi-Team Systems

Several chapters within this textbook underscore the key role that teamwork and coordination play in cybersecurity operations. Zaccaro, Luckes, and Zakzewski suggest that failures often stem not from technological shortcomings but rather poor teamwork and team orchestration. Their Input-Process-Output model provides a helpful framework from which to view cybersecurity incident response teams, their function, and how they may be improved. Similarly, Berger, Khalid, Linhardt, Fernández Castillo, and Salas develop this by focusing specifically on SOCs, with inputs, mediating processes, and outputs. This work highlights factors such as shared mental models, psychological safety, trust, and workload management that contribute to team outcomes. Brennan, Fausett, and Lazzara look at a specific kind of team process: the handoff, or transfer of responsibility from one provider to another. Together, these chapters establish teamwork as a necessary but often ignored dimension of cybersecurity excellence, requiring attention and development rather than emerging spontaneously from technical expertise. After all, a team of experts does not make an expert team (Reyes & Salas, 2019).

Organizational Resilience and Learning

Resilience emerges as a key theme across multiple chapters. Boskeljon-Horst, Dekker, and Dekker highlight four essential capacities that contribute to organizational resilience: monitoring, responding, learning, and anticipating. These capabilities enable cybersecurity organizations to adapt to evolving threats and recover effectively from incidents. The authors advocate for *restorative* rather than *retributive* approaches when incidents occur. This perspective recognizes that punitive responses to cybersecurity incidents often drive problems underground and inhibit the transparency needed for systemic improvement. Berger, Khalid, Linhardt, Fernández Castillo, and Salas similarly highlight adaptability and resilience as key outputs of effective SOC teams, emphasizing the importance of learning processes and stress management strategies. Their work suggests that team resilience depends not only on technical capabilities but also on psychological factors like team climate and leadership.

Innovative Defensive Approaches

The textbook also introduces novel approaches that extend beyond traditional defensive postures. Gutzwiller, Lewis, Pharmer, Clegg, Aggarwal, Henery, Ray, and Martey's chapter on oppositional human factors represents a particularly innovative contribution, proposing that defenders can strategically exploit the cognitive limitations and biases of attackers. By deliberately introducing friction and triggering decision-making biases, defensive systems might become more resistant to human attackers. Similarly, Simonson, Foquet, and Fausett explore how a system survivability framework can strengthen healthcare organizations against social engineering attacks, advocating for a shift from reactive to proactive cybersecurity measures. These innovative approaches demonstrate that human factors research can contribute not only to reducing defensive vulnerabilities but also to creating new defensive advantages. By understanding the psychological dimensions of both defense and attack, cybersecurity professionals can develop more sophisticated and effective protective measures.

Domain-Specific Applications of Human Factors in Cybersecurity

Cybersecurity concerns span across domains. Human factors expertise provides crucial insights for cybersecurity through methods like usability testing, cognitive task analysis, and naturalistic decision-making. This intersection of human factors and cybersecurity becomes vital in high stakes environments where cyberattacks can lead to devastating consequences, such as healthcare, aviation, and transportation. Addressing these challenges requires an interdisciplinary approach that integrates human factors specialists, cybersecurity experts, and domain practitioners. While the principles discussed in this book apply broadly, specific operational requirements, regulatory environments, and threat landscapes will vary across sectors. Implementation of the principles herein should be tailored to these specific operational requirements.

CHAPTER OVERVIEW

Applying Lessons Learned from the Safety of Sociotechnical Systems to the Human Factors of Cybersecurity

A goal of this text is to provide the reader with the current state of human factors science as it relates to cybersecurity. Following this introductory chapter, *Applying Sociotechnical Systems Safety to Cybersecurity* by Schuster explores the application of sociotechnical system theories to cybersecurity. In this chapter, Schuster demonstrates that addressing cybersecurity challenges necessitates an interdisciplinary approach that centers human factors. The author draws parallels between cybersecurity and other high-risk domains such as aviation and healthcare, demonstrating how frameworks like SEIPS (Systems Engineering Initiative for Patient Safety) and CRM (Crew Resource Management) could be adopted to improve cybersecurity practice. A three-pronged research agenda is presented: (1) applying existing safety frameworks to cybersecurity contexts, (2) conducting cognitive task analysis to understand the unique thinking processes of cybersecurity professionals, and (3) developing validated measures through quantitative research. This chapter advocates for an approach to cybersecurity that treats people as essential partners in security efforts rather than sources of risk.

Human Factors within the Cybersecurity Domain

The chapter *Human Factors within the Cybersecurity Domain* by Boskeljon-Horst, Dekker, and Dekker begins with a discussion of human error, presenting a view that establishes human error not as a cause of issues, but rather as a symptom of deeper systemic problems. The authors draw on human factors research to show how errors connect to features of people's tasks, tools, and operational environments rather than individual failings. The authors discuss schools of thought around human error, detailing an evolution from traditional error prevention approaches to understanding organizational resilience. The authors mention four key capacities: monitoring, responding, learning, and anticipating. Cybersecurity organizations may often misattribute problems to "human error" instead of examining systemic factors. In response, the authors advocate for a restorative justice approach rather than retributive models when incidents occur. The restorative approach focuses on repairing harm, moral engagement, worker reintegration, and organizational learning. Human error is inevitable, and organizations should create climates that enhance learning rather than attributing blame – a perspective that could change how you view your own organization's approach to mistakes and learning.

Cybersecurity Incident Response Teams and Multiteam Systems: An Input-Process-Output Framework

In *Cybersecurity Incident Response Teams and Multiteam Systems: An Input-Process-Output Framework*, Zaccaro, Luckes, and Zakzewski argue that cybersecurity failures often stem from poor teamwork and coordination, rather than technical shortcomings. To augment this, the authors present an Input-Process-Output model to better understand cybersecurity incident response, explaining how teams and multi-team systems (MTSs) operate in this context. The authors discuss structural elements of these teams, such as goal hierarchies and interdependencies, and identify key team processes such as proactive planning and reactive responses. Two major challenges are identified, (1) achieving social maturity (the development of collaborative skills) and, (2) managing differentiation among component teams in an MTS. The authors emphasize that effective incident response requires not only technical expertise but also strong teamwork capabilities, especially where multiple specialized teams must intensely collaborate across organizational boundaries.

Enhancing Cybersecurity through Teamwork: Developing Effective Cybersecurity Operation Center Teams

In *Enhancing Cybersecurity through Teamwork: Developing Effective Cybersecurity Operation Center Teams*, Berger, Khalid, Linhardt, Fernández Castillo, and Salas propose a framework to help us better understand what team effectiveness means in cybersecurity operations centers (SOCs), followed by a discussion of learning, adaptability, and resilience. The authors introduce key inputs, such as the SOC analysts, the SOC team, and organizational context. Next, mediating mechanisms like decision-making, shared mental models, psychological safety, trust, and workload are described. These factors thus influence key outputs, including team effectiveness, adaptability through briefing and learning, and resilience via stress management strategies. These components describe how an SOC team can function efficiently and prioritize well-being in high-pressure environments.

Insights and Recommendations for Improving Handoff Processes in Cybersecurity Operations Centers

The chapter *Insights and Recommendations for Improving Handoff Processes in Cybersecurity Operations Centers* by Brennan, Fausett, and Lazzara explores the critical but understudied area of handoff processes in SOCs. The authors describe that effective handoffs are essential for maintaining continuous cybersecurity operations, particularly as SOCs evolve to include distributed, global teams working across time-zones. The authors draw on research from high-reliability domains like healthcare, aviation, and nuclear power to examine how failures in handoff communication can lead to catastrophe. The authors also mention unique challenges facing SOC handoff operations, including (1) reliance on event management systems, (2) cultural differences in global operations, (3) confidentiality, integrity, and availability of information. Next, recommended practices are presented such as standardization, the explicit transfer of responsibility, closed-loop communication, and the development of shared mental models. The chapter then investigates the SOC SEAT-SWAP handoff mnemonic (Staff, Explanation, Awareness, Transition, Situation, Written, Accurate, Persistent), illustrating best handoff practices in application through a verbal handoff scenario involving a credential surfing attack. This chapter provides guidance to practitioners and future directions for research in this important area of information transfer.

Oppositional Human Factors

The chapter *Oppositional Human Factors* by Gutzwiller, Lewis, Pharmer, Clegg, Aggarwal, Henery, Ray, and Martey introduces a new way of thinking about cybersecurity defense that focuses on understanding and then exploiting the cognitive limitations and capabilities of attackers. Gutzwiller et al.'s chapter serves as an important reminder that cyber attackers are also *human*. This means that despite their technical prowess, they remain susceptible to the same cognitive biases and attentional limitations as the rest of us. As a result, these researchers developed the oppositional human factors framework which in essence reverses traditional human factors principles. Instead of using human factors knowledge to improve systems and make them more user-friendly, they deliberately introduce friction through elements that trigger decision-making biases and attentional disruptions in attackers. The authors discuss the results of studies in this area investigating how biases like sunk cost fallacy, confirmation bias, asymmetric dominance, and anchoring can be implemented, strategically, in defensive systems. This chapter concludes that oppositional human factors represents a promising psychological complement to the technical defenses, potentially a more enduring one as cognitive vulnerabilities are much harder to overcome than technical ones.

Complex Healthcare Systems in the Modern Social Engineering Landscape: The Human Factors of System Survivability

In *Complex Healthcare Systems in the Modern Social Engineering Landscape: The Human Factors of System Survivability*, authors Simonson, Foquet, and Fausett explore how social media can amplify social engineering vulnerabilities particularly within healthcare environments. Healthcare professionals may be particularly susceptible to social engineering attacks due to the high-stress, high workload environments in which they operate. The chapter discusses the system survivability framework (prevention, mitigation, and recovery) as a way to strengthen healthcare against social engineering attacks. Authors advocate for a shift away from reactive toward proactive cybersecurity measures and the development of organizational cultures that support security awareness.

A Survey of Machine Learning for the Detection and Identification of Internet of Things (IoT) Devices with Application in Aviation

Machine Learning for Device Identification with Application in Aviation Cybersecurity by Liu, Wang, and Song provide a comprehensive overview of machine learning techniques for identifying rogue IoT devices in increasingly connected environments. The authors examine both cryptographic and non-cryptographic approaches, focusing on physical-layer identification that leverages unique hardware imperfections to create distinctive "fingerprints" for wireless transmitters. Throughout the text, the authors give particular attention to human-machine interface implications, emphasizing how each approach affects user experience and decision-making.

Cybersecurity and Ground Transportation

The chapter *Cybersecurity and Ground Transportation* by Xiaojian Jin explores the complex cybersecurity challenges facing modern transportation. Specifically, this chapter investigates the incorporation of automated driving systems to commercial vehicle fleets. The transition to mixed fleets, with both traditional and automated driving systems introduces new vulnerabilities: software vulnerabilities, data protection concerns, physical security risks, and human factors considerations. The author highlights that cybersecurity in transportation is not merely a technical issue but one that requires an integrated approach combining human-centered design principles with robust technical protections.

CONCLUSION

Creating a path forward for cybersecurity doesn't necessitate perfect technology or perfect humans. It does, however, require the thoughtful integration of both into systems that recognize the complex realities of our digital world. It is our hope that this textbook will serve as a foundation for this integration, and inspire continued research and innovation at the convergence of human factors and cybersecurity.

REFERENCES

Cafazzo, J. A., & St-Cyr, O. (2012). From discovery to design: The evolution of human factors in healthcare. *Healthcare Quarterly*, *15*, 24–29.

Goh, P. (2021). Humans as the weakest link in maintaining cybersecurity: Building cyber resilience in humans. In M. Khader, W. X. T. Chai, & L. S. Neo (Eds.), *Introduction to cyber forensic psychology: Understanding the mind of the cyber deviant perpetrators* (pp. 287–305). World Scientific Publishing.

Jain, A. K., & Duin, R. P. (2004). Introduction to pattern recognition. In RL Gregory (Ed.), *The Oxford companion to the mind* (2nd ed., 698–703). Oxford University Press.

Keebler, J. R., Lazzara, E. H., Wilson, K. A., & Blickensderfer, E. L. (Eds.) (2022). *Human factors in aviation* (3rd ed). Elsevier.

Pujari, S. R., & Hussain, M. (2024). Human factor in cybersecurity: Behavioral insights into phishing and social engineering attacks. *Nanotechnology Perceptions*, *20*, 630–642. https://doi.org/10.62441/nano-ntp.vi.3556

Reyes, D. L., & Salas, E. (2019). What makes a team of experts an expert team. In R. F. Subotnik, P. Olszewski-Kubilius, & F. C. Worrell (Eds.), *The psychology of high performance: Developing human potential into domain-specific talent* (pp. 141–159). American Psychological Association.

Stone, N. J., Chaparro, A., Keebler, J. R., Chaparro, B. S., & McConnell, D. S. (2017). *Introduction to human factors: Applying psychology to design*. CRC Press.

Rooparaghunath, R. H., Harikrishnan, T. S., & Gupta, D. (2020). Trenchcoat: Human-computable hashing algorithms for password generation. In S. Krenn, H. Shulman, & S. Vaudenay (Eds.), *Cryptology and network security* (pp. 167–187). Springer. https://doi.org/10.1007/978-3-030-65411-5_9

Yan, Z., Robertson, T., Yan, R., Park, S. Y., Bordoff, S., Chen, Q., & Sprissler, E. (2018). Finding the weakest links in the weakest link: How well do undergraduate students make cybersecurity judgment? *Computers in Human Behavior*, *84*, 375–382.

1 Applying Sociotechnical Systems Safety to Cybersecurity

David Schuster

INTRODUCTION

The Cybersecurity Problem

Ensuring the cybersecurity of organizations is a massive challenge. One driver of the impact of cybersecurity risk is the scale of the modern Internet, which interconnects nearly all organizations globally and serves over five billion users (Statista, 2024). Cybersecurity, which can be defined as the maintenance of the confidentiality, integrity, and availability of information systems (Fruhlinger, 2024), is also an evolving and highly interdisciplinary problem requiring interdisciplinary solutions. Focusing on the problem of cybersecurity risks to organizations, especially businesses, does little to make the problem more manageable because of the interconnected nature of the Internet. Users regularly access the Internet from a variety of devices to engage in an enormous variety of tasks. Human actions that can harm or support an organization's cybersecurity can potentially come from any other user of the Internet.

The complex interconnection of organizations creates a problem with both large financial and human resource implications. As one example, UnitedHealth Group reported $872 million loss from a ransomware attack in 2024. A payment platform used by providers, insurers, and other healthcare professionals was targeted. Upon discovery of the ransomware attack, UnitedHealth shut down the payment platform. This was disruptive to pharmacy services, medical claims, patients, and providers. This is a large loss expressed in dollars, but the example also illustrates the human implications; in a society dependent on confidentiality, integrity, and access to information systems for critical tasks, the fallout from cyberattacks goes beyond inconvenience to affect the health, lives, and safety of millions of people. While corporate losses are often large and easier to calculate, the human costs are less visible. Behind the dollar amounts are substantial risks to all participants in an information society.

As another example, nearly a third of US drinking water systems sampled by the US Environmental Protection Agency (EPA) were vulnerable to cyberattack (Vigliarolo, 2024). The EPA has not described these vulnerabilities nor how they could be mitigated. Further, their sample only included the largest water systems serving more than 50,000 people. No such data exists for smaller water systems. This analysis is troubling because it implies great risk without providing any information about whether or how changes can be propagated to protect people affected. It further illustrates the interconnected nature of cybersecurity risk; people do not need

DOI: 10.1201/9781003413707-2

to participate to be vulnerable. It also underscores that threats to cybersecurity are often threats to safety.

The Need for Human Performance Development

While the effects of cyberattacks can reverberate to affect large populations of people, the causes of cyberattacks are also diverse. For example, Verizon reported that 68% of data breaches involved what they called the human element (2024). People are frequently cited as the weak link in cybersecurity (Zimmermann et al., 2024). While these observations begin to hint at the interdisciplinary nature of the problem, a human-centered perspective is needed. A human-centered perspective focuses on the goals of people and leverages human decision-making and performance to improve security outcomes. This is necessary because threats and errors are inevitable, and a system of cybersecurity that depends on perfect human behavior to avoid catastrophe will be brittle. Similarly, a strategy, such as training, that aims to perfect human behavior is unlikely to be effective. Unless the impacts of human behavior can be completely engineered out, cybersecurity will continue to function as a sociotechnical system in which the operation of many people and machines interact to affect outcomes (Schuster & Keebler, 2019).

The continued need for human performance is evidenced by demand for cybersecurity professionals. The workforce does not have enough cybersecurity professionals who continue to be a critical resource for organizations. For example, CyberSeek, a cybersecurity workforce tracker built in partnership between industry and the US government, tracks unfilled cybersecurity positions in the United States. For the past few years, the unmet workforce need has hovered around half a million workers (CyberSeek, 2024). As of this writing, CyberSeek reports that there are enough cybersecurity workers to meet 83% of the demand. Clearly, there is a great need for further development of training pipelines to prepare for these careers. While a great deal of funding and effort is applied to solving cybersecurity problems, we still lack understanding of the cognitive aspects of the work. With better understanding of the knowledge, skills, and attitudes that predict success in cybersecurity careers, training and career development pipelines could be further developed.

Purpose of This Chapter

Fortunately, there are lessons that can be learned from other domains. The central argument in this chapter is that the interdisciplinary cybersecurity problem could be better addressed with a greater incorporation of the tools, research methods, and theory of sociotechnical systems currently helping other domains, like healthcare systems. Sociotechnical systems are complex, high-risk domains that bring together people and technology (Holden et al., 2013). Key features of sociotechnical systems, including information systems, are what Vicente (1999) called *dimensions of complexity*, which include uncertain data, large problem spaces, and heterogeneous perspectives of the agents that affect outcomes. A recent report from the US Office of the National Cyber Director (2024) described several long-standing cybersecurity challenges. These challenges seemed somewhat intractable despite rapid technological evolution. They illustrate why a purely technological approach is insufficient to address cybersecurity at scale. Table 1.1 links the cybersecurity challenges to

TABLE 1.1
Mapping of Long-standing Cybersecurity Challenges (Office of the National Cyber Director, 2024) to Dimensions of Complexity (Vicente, 1999)

Challenge	Explanation	Application
Office of the National Cyber Director (2024)		To Vicente's sociotechnical system complexity (1999)
Structural asymmetries	Defense is relatively costly, and attack is relatively cheap. Defenders must win every time, but attackers only need to win once.	**Hazard** means that the prevention of catastrophe depends on human behavior, especially in abnormal circumstances like during an attack.
Legacy protocols and architectures	Foundational technologies are old and not easily updated, even if they are vulnerable	**Coupling** means that subsystems interact to affect risk, making it hard to forecast and address. For example, old versions of the TLS encryption protocol are vulnerable, but software and devices must be updated to support newer versions, or the participation of old versions must be supported.
Emerging digital technologies	New tools make attacks easier or may give a disruptive, substantial advantage to attackers.	**Automation** is a disruptive feature because it enables new capability imperfectly; people are still needed to collaborate with automation. For example, automation of routine datacenter tasks may lead to false alarms, requiring human investigation.
Third-party service providers	Risks affect both an organization's infrastructure and the organization's vendors, who also serve other organizations.	**Distributed** agents affect outcomes but do not work for one organization, reducing visibility and control. For example, a backdoor secretly embedded in a microprocessor could lead to vulnerability in many products that incorporate it.
Private ownership and operation	The Internet is a public service that is dependent on private organizations, including for space-based communication.	**Social** systems depend on effective teamwork and multiteam coordination, which can include the collaborative work of thousands of people. For example, a large company discovering a threat that could affect the public may be disincentivized from disclosing it.
Cyber workforce	The pipeline of cybersecurity professionals is insufficient to meet demand.	**Dynamic** systems have lag; interventions to improve the cyber workforce have slow impacts, making it challenging to understand what works. For example, it is difficult to evaluate the effects of cyber workforce development efforts.

dimensions of complexity (Vicente, 1999) to illustrate that, from a sociotechnical systems perspective, these are challenges that have been known for over 25 years, and they have been investigated and addressed in other domains.

The purpose of this chapter is to argue for a sociotechnical systems approach to strengthening cybersecurity in organizations, both for immediate use in organizations and as a framework for conducting mixed-methods research. In subsequent sections, I offer examples of tools from other sociotechnical systems and provide examples for their application to cybersecurity. With this chapter, I provide a research agenda so that more practitioners and scientists can enter the field and collaborate to address this large and important problem.

THEORETICAL FRAMEWORK

Toward an Interdisciplinary Science

Having argued that a systems perspective is needed, I next describe the need for an interdisciplinary science of cybersecurity. Such a science would investigate and understand how cybersecurity solutions operate within a sociotechnical system and inform all aspects of cybersecurity practice with evidence-based guidance, including in the design of tools, measurement of outcomes, the development of training, and strategies for workforce recruitment and selection. What makes this perspective particularly valuable for cybersecurity is that it acknowledges the complex interdependencies caused by systems of people and technology operating at scale.

Value of Human Factors/Ergonomics

Human factors/ergonomics (HF/E) is a field that is well-positioned to apply a sociotechnical systems perspective to improve safety and performance in information systems. HF/E offers a diverse toolbox to understand and diagnose human performance in work contexts. This toolbox includes methods for practitioners to diagnose and improve sociotechnical system performance. This toolbox is also rich with methods that researchers can use to measure and identify factors affecting sociotechnical system performance, leading to methods specifically tailored to and validated in cybersecurity contexts. One especially powerful method of HF/E is the capability to understand the cognitive aspects of work.

Human participation in cybersecurity may share features of other domains of successful inquiry to HF/E, including aviation, military command and control, and healthcare. These domains are diverse in their tasks, human participants, and environmental context. At the same time, we understand that complex interactions among people, technology, tools, and environmental factors affect outcomes and are often under-considered. To investigate these, HF/E provides empirical methods to investigate human aspects of work that can be domain agnostic to an extent. This is possible in an investigation into the cognitive aspects of work, and the insights it generates have the potential to apply to the new domain. That is, some principles of human perception and cognition may apply similarly to airplane flight as they do in an operating room, as one example. To pick the low hanging fruit, one needs to understand the commonalities across domains while discovering what is unique in the new domain. While this is taking place, more substantive empirical research can be spun up.

In studying this problem, I have been surprised by the level of unmet need for evidence-based cybersecurity solutions from both HF/E and the nascent interdisciplinary science of cybersecurity. This may be because of some of the idiosyncrasies of cybersecurity. For one, its scale is a level of magnitude above other domains of interest because it intertwines all of them. A cybersecurity threat could lead to devastating outcomes in aviation, military, and healthcare environments. The speed at which the domain evolves is also unprecedented; a cat-and-mouse game between attackers and defenders speeds the evolution of the state-of-the-art. People do not evolve in this way, and the research infrastructure has not been able to keep pace with this rapid evolution. I propose that the solutions may lie in the participation in an interdisciplinary science of cybersecurity that leverages diverse methods of HF/E research and practice to center people and their goals. In other words, the technical side of cybersecurity needs to become more interdisciplinary by incorporating a sociotechnical systems perspective; the HF/E aspect of cybersecurity needs to move faster and with greater participation.

Proposed Approach

With the goal of promoting participation in HF/E in cybersecurity and interdisciplinary cybersecurity generally, I offer a methodological approach that covers several perspectives. I also describe how these perspectives are complementary. Progress in cyber will require more HF/E researchers to enter the domain and may require HF/E researchers to broaden their approach. For example, laboratory experimentalists offer value by isolating variables of interest and developing and validating measures of relevant human behavior, but researchers and practitioners in the field are essential collaborators to understand how such variables operate in the complex and rapidly evolving context.

The methodological approach presented here provides a comprehensive approach to conducting human-centered research. It builds upon what Zimmermann et al. called enabling approaches, which "treat humans as a resource and a partner in security efforts" (2024, p. 73). Zimmermann et al. distinguished enabling approaches from two other categories: *constricting approaches*, which view users as an enemy, and *constraining approaches*, which merely tolerate the behavior of people, who are viewed as a source of risk (Zimmermann et al., 2024). Although they discussed enabling approaches in the context of user behavior across the organization, it is especially applicable to cybersecurity professionals within organizations. The work of cybersecurity professionals is insufficiently understood, leading to potentially inefficient application of human performance. There is no room for such inefficiency given the workforce shortage. Put another way, a more comprehensive solution to cybersecurity problems requires full application of the unique talents of people. Without this understanding, any approach to engineer people out through automation may appear necessary but is unlikely to be effective.

The proposed approach is described in Table 1.2. Proposed research can advance along three fronts. The first is in leveraging the basic principles of

TABLE 1.2
Overview of Proposed Approach

Approach	Time to Actionable Recommendations
1. Apply sociotechnical system safety frameworks to security	Days to weeks
2. Conduct task analysis to gain insight into cognitive performance	Weeks to months
3. Validate measures and recommendations using quantitative and mixed-methods research	Months to years

human-machine interaction by understanding how cognition in other, better-understood systems such as aviation and healthcare, respond to similar challenges of complexity present in sociotechnical systems. This can be achieved through adaptation of existing models used in those systems and the adoption of tools and research methods. I will focus on two specific examples; the first is the SEIPS 101 toolkit (Holden & Carayon, 2021), which is based on the Systems Engineering Initiative for Patient Safety (SEIPS) 2.0 model (Holden et al., 2013). The second is to improve teamwork by applying crew resource management (CRM; Helmreich, 2000a). Integrating these approaches has immediate implications for practice, so researchers and practitioners can get started improving systems and centering the goals of people immediately. This process should reveal where these methods are suboptimal to cybersecurity, and in doing so, will start to reveal unique characteristics of cognition in cybersecurity contexts. The benefit of doing this process and learning from misapplication is the development of research questions to be investigated along the second front, which involves qualitative research. At the same time, people evolve slowly, so there will be lessons that have been learned in other domains that remain constant even when the technological context changes. For example, we better understand how otherwise safe systems can lead to unsafe operation when normal conditions leave professionals under-prepared for anomalous circumstances (Wickens, 2002). Parallels can be drawn to cybersecurity, where the fast evolution of attack methods leads to signals and outcomes that were previously unknown.

By applying models and methods from other domains, limitations of those models will emerge. An important parallel task is the identification of critical factors affecting and affected by human cognition. The second front is to conduct qualitative research. Specifically, methods of cognitive task analysis can reveal cognitive fundamentals of the work by learning what makes cybersecurity professionals effective. In this research, the practitioner is the key to understanding the domain and is a partner in the work. Such research embraces complexity and constraint present in the field. According to Creswell and Creswell (2018), qualitative research offers value in identifying variables of interest. While cybersecurity is not a new problem, rapid technological evolution can have the effect of shifting the variables of interest, such as when processes formerly done by people are automated.

Quantitative research is necessary to establish measures of key constructs and, in time, confirm cause-and-effect relationships. These methods reduce complexity though isolating variables of interest. Shadish et al. (2001) described three aims of psychological research: *causal description*, *causal explanation*, and *causal generalization*. *Causal description* is the identification of cause-and-effect relationships. Essentially, it is the outcome of quantitative research, especially randomized experimentation. It could identify, for example, that a cybersecurity career awareness campaign causes young adults to report greater interest in cybersecurity careers. *Causal explanation* describes the mechanisms of such relationships. It involves the development of theory that is facilitated by qualitative research. For example, interviews with young adults to describe why they are or are not interested in cybersecurity careers could, among other insights, explain what ingredients are essential for cybersecurity career awareness content. *Causal generalization* identifies the circumstances under which such a relationship exists, which is possible with continued mixed-methods research. To continue this example, it may be the case that the factors that motivated young adults in 2019 to pursue cybersecurity careers differ from young adults in 2026 due to a cohort effect. Continued research offers value to understand the boundaries of such effects.

In all, the remainder of this chapter offers a high-level roadmap for overlapping human-centered research activities. Practitioners will find immediately applicable, though imperfect, tools for analyzing elements of their sociotechnical system. Researchers will discover both quantitative and qualitative methods for studying cybersecurity work, starting with discovery of the strengths that people bring to cybersecurity work. For those experienced in HF/E but new to the cybersecurity domain, I hope to accelerate entry into this field by drawing parallels to domains that readers may already work in. By conducting this work, researchers and practitioners can ensure that cybersecurity challenges are responded to with every available resource, including the unique talents of people.

RECOMMENDATIONS FOR PRACTICE

Next, I argue for the value of lessons learned from cognitive engineering in other systems that may be applicable to cybersecurity systems. To do this, I develop two examples. First, I provide examples from the design of medical systems. Next, I describe teamwork interventions used in aviation that are underutilized in cybersecurity teams. The focus in this section is on tools for practice that have the fastest route to application.

Recommendation 1: Improve Current Systems by Applying Safety Frameworks to Cybersecurity

The SEIPS 2.0 model (Holden et al., 2013) is a framework developed to apply sociotechnical systems thinking to the assessment, design, and improvement of healthcare systems. In prior work, we argued that the SEIPS 2.0 model could be applied to cybersecurity and adapted the concepts of the model to apply to an organizational

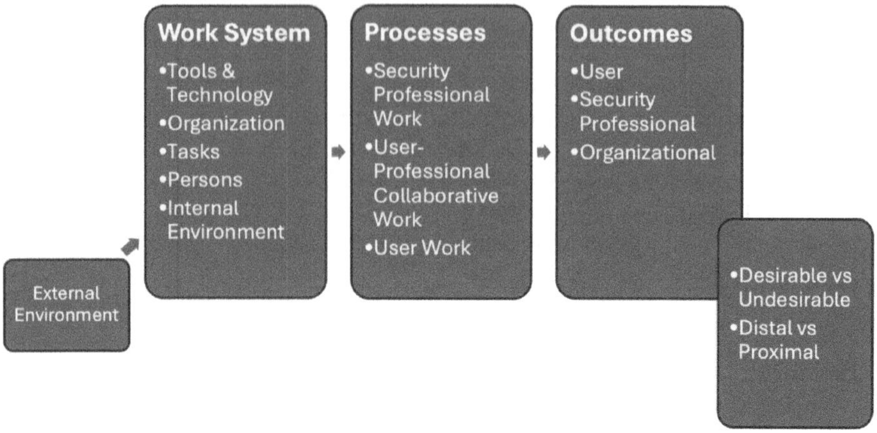

FIGURE 1.1 Schuster & Keebler's (2019) adaption of Holden et al.'s SEIPS 2.0 model (2013).

cybersecurity context (Schuster & Keebler, 2019; see Figure 1.1). The SEIPS 2.0 framework highlights interactions between work system elements, especially patients and clinicians. For cybersecurity professionals, this can be expanded to consider interactions between cybersecurity professionals and users. Common to healthcare and cybersecurity is the relevance of the work system elements, including tools, the organization, tasks, persons, and the internal and external environment. Although the model appears linear, it suggests complex interactive effects between work system elements, processes, and outcomes. In other words, elements of the work system drive processes involving those elements, and resulting outcomes also affect processes and design of work system elements. SEIPS 2.0 has benefited from 15 years of development, including 7 straightforward SEIPS tools that can be easily adapted to a cybersecurity context. These tools and suggestions for their use are described in Table 1.3.

These tools may be of interest to both practitioners and researchers, as they result in immediate actionable insights while also uncovering aspects for more research. The PETT scan can be used as an analysis in the planning of IT projects while considering the cybersecurity impact. We have observed that even enterprise corporations operating at the state-of-the-art face barriers to implementing strategies such as teamwork (Schuster et al., 2024). The PETT scan would allow practitioners to better anticipate the success of a project due to factors that might not be considered in a more limited project plan. In essence, it allows for a more comprehensive project plan, including the effects of the broader work system while focusing on security goals.

In medical settings, the people map can generate insight into how roles interact, and thus hint about opportunities for gains through teamwork, independent of organizational structures, which may be idealized and hierarchical. Cybersecurity professionals use a multitude of tools in performance of their work, but efficiencies may exist that allow organizations to respond to threats faster if gaps in what the tools offer are identified. At a time where artificial intelligence (AI) is enabling new capabilities, identification of a targeted need for automated help matched to specific supportive tasks could lead to new human-centered tools. The tasks and tools matrix provides this insight. Another use for the SEIPS 101 tools could be to help to develop

TABLE 1.3
Application of the SEIPS 101 Tools to Cybersecurity Work

SEIPS 101 Tool from Holden and Carayon (2021)	Purpose	Potential Insights
PETT scan	Identifies barriers and facilitators across work system factors	Understanding of holistic impact of security interventions across the organization; predict barriers to cybersecurity interventions to ensure cyber projects are successful
People map	Goes beyond an idealized organizational chart to show how roles relate in the operational environment	Development of personas, job descriptions that more accurately reflect the work being performed; description of people involved in cybersecurity efforts and how their work is related; methods of more efficient teamwork or allocation of resources
Tasks and tools matrix	Describes the tools used for each task	Identification of task needs not provided by current tools; understanding of how human analysts are needed; development of new tools; refined playbooks
Outcomes matrix	Inventory of desirable and undesirable proximal and distal outcomes	Refined metrics; understanding of organizational impact of security outcomes
Journey map	Represents how stakeholders interact with other system elements	Understanding of a process over time in relation to other system elements; refined playbooks; models of user or stakeholder behavior; identification of barriers to user compliance
Interactions diagram	Visually represents how work system factors interact	Sensemaking during incident response that centers organizational impact; understanding of remediation impact
Systems story	Tells a compelling story about how the work system factors interact to affect an outcome	Training for cybersecurity awareness; understanding of how cybersecurity is a shared responsibility; advocacy for cybersecurity needs

and improve playbooks, which are models used to manage cybersecurity incidents (Onwubiko & Ouazzane, 2022). Specifically, the insights of these methods may help in preparation for incident response and make the incident response more effective by making plans more realistic.

Recommendation 2: Develop Cyber Professional Teamwork by Leveraging Crew Resource Management

Constricting approaches (Zimmermann et al., 2024) that frame human behavior as a major source of risk support a cycle of blame and underutilization of human capabilities. Analysis of incidents from the perspective of locating and eliminating human error can perpetuate blame. Aviation is a comparative domain that has struggled with the issue of human error but made substantial progress by adopting a sociotechnical systems approach. Seeing people as the weak link and focusing on user error results in an underutilization of available resources. This is counter-productive because human decision-making remains essential. A human-centered approach would recognize that cybersecurity professionals are a critical resource in short supply. Therefore, their knowledge and skills need to be maximally applied to the high-stakes problems at hand. A symptom of suboptimal utilization of knowledge and skills of people is an underutilization of teamwork (Gutzwiller et al., 2016; Schuster et al., 2024), despite recommendations for the design of teamwork in cybersecurity being available for some time (e.g., Tetrick et al., 2016).

Next, I propose parallels to aviation and one specific, evidence-based practice that has improved safety, CRM. CRM is in its seventh major revision (Terzioğlu, 2024). It incorporates a threat and error management framework (Helmreich et al., 1999). In the threat and error management model (Helmreich, 2000a), internal and external sources of risk and human error are examples of threats. Threats, which can negatively impact safety, are unavoidable. Examples of common threats in aviation include terrain and adverse weather (Helmreich, 2000b). The threat and error management model fits a sociotechnical systems approach because it models interactions that can lead to recovery or an incident. Two sources of mitigation are the focus of CRM. The first is the recognition of threats and the avoidance of error. The second is the detection of, and response to, error. Failure of error detection and response is, itself, a form of error and can compromise safety.

Together, threat and error management, and CRM as a training framework, shifts from individual blame to understanding the interconnected nature of risk in sociotechnical systems. Verizon identified error as a factor in 28% of breaches (2024), and an earlier analysis suggested human error contributed to 50% of cases (PWC as cited by Evans et al., 2016). This is in line with aviation, where Terzioğlu (2024) estimated that crew errors contributed to 70% of aviation accidents. As with accident investigation, analysis of successful breaches provides a biased sample of incidents with the worst outcomes. Threats and error that contribute to risk, but are not associated with published breaches, are not visible.

The goal of CRM is to recognize how sociotechnical system elements affect safety, and to empower all human participants to contribute to safety. CRM is not a single form of training, but an approach to training that has been shown to improve team performance (Pizzi et al., 2001). One approach to evaluating the effectiveness of CRM is to measure the change in safety attitudes using the Flight Management Attitude Questionnaire (FMAQ; Kanki et al., 2010). This measure has five dimensions: command responsibility, my stress, stress of others, rules and order, and communication. Each of these can be applied to cybersecurity teams, with examples provided in Table 1.4.

Because security depends on cybersecurity professionals who are in short supply, their work is critical and their individual and collective performance is worthy of study and improvement. CRM was developed to improve the safety of critical, high performing systems by training for more effective teamwork and attitudes that

TABLE 1.4
Application of CRM Dimensions to Cybersecurity Teams

Dimension of the FMAQ (Kanki et al., 2010)	Attitudes and Behaviors in Aviation	Application to Cybersecurity Teams
Command responsibility	Legally, the pilot is ultimately responsible for safety, and co-pilots are also responsible for the safety of the flight	All cybersecurity professionals are responsible for security.
My stress	Workload, physical stressors, and psychosocial stressors can negatively impact safety	Cybersecurity professionals need to develop awareness of their own stress and fatigue levels.
Stress of others	Team members may have diminished capacity in an emergency due to stress	Stress and fatigue in others are likely during incident response and contribute to risk, so they need to be recognized.
Rules and order	Rules and procedures for safe and effective flight (e.g., following checklists) are developed from experience and need to be followed	Checklists, playbooks, and procedures are derived from experience and need to be used consistently.
Communication	The captain is responsible for leading communication and coordination	Managers contribute to security by coordinating teams and supporting effective communication; Effective coordination is necessary for effectiveness.

support more effective teamwork. In cybersecurity teams, these attitudes can be trained and assessed as a first step.

This section provided an overview of two frameworks from sociotechnical systems outside of cybersecurity and offered suggestions for the application of their tools to the work of cybersecurity professionals. This aspect of the approach is likely to be worthwhile because of the commonalities across sociotechnical systems, and the application of these frameworks will improve understanding of how sociotechnical system elements in cybersecurity are unique. Further, it will help develop scientific understanding of security and safety.

PROPOSED RESEARCH AGENDA

Recommendation 3: Conduct Task Analysis to Gain Insight into Cognitive Performance

Qualitative research involves inductive reasoning; the deep analysis of specific examples leads to insights, which are generalizations from the observed case. Cognitive task analysis is a collection of qualitative research methods that lead to models of aspects of the sociotechnical system. Cognitive task analysis specifically allows for exploration of the people elements and how they perform cognitive work (Klein & Wright, 2016). Cognitive task analysis can help identify macrocognitive processes that are the intermediary between low-level cognitive resources and performance outcomes (Schuster, 2020).

So far, published CTAs have supported the importance of human cognition and performance in the domain. One advantage is that by providing workflow models, CTAs can highlight what cybersecurity professionals contribute and the need for human decision-making. In turn, this can be helpful to design human-centered automation (Trent et al., 2019).

CTA has been applied to cybersecurity but is plagued by several challenges. For one, CTAs require insider access that organizations are reluctant, or sometimes legally prohibited, to share. A notable exception to this is work by Ofte (2024), who was able to conduct a goal-directed task analysis in an industry security operations center (SOC) because of his role as an employee. In this work, four high-level goals of the SOC were described: monitor, detect, and escalate incidents; mitigate incidents; determine cause of incident; and re-establish secure system operation. Network incidents and security incidents differ in that network incidents require identification of cause before mitigation; security incidents require mitigation to be prioritized while developing SA because investigation requires a safe setting. Security and operation sometimes conflict; teams must balance integrity and availability. Insights such as these show how rich and useful the data are when researchers are given access, and Ofte (2024) has demonstrated that publication of insights can be done without revealing sensitive data. Previously, we have argued for a data collection strategy based on elicitation by critiquing (EBC; Scott et al., 2018). Based on the work of Miller et al. (2006), EBC directs the participant to review a novice's performance of a task. The knowledge elicited comes from the participant's description of how their approach to the scenario would differ or match that of the novice. Two aspects of this technique have been especially useful for us in dealing with the problem of limited access. First, the technique is not dependent on participants discussing details of actual incidents encountered in their work. Second, because the stimuli in the research consist of a scenario and a necessarily imperfect handling of the scenario, the stimuli can be developed at a low cost. That is, cybersecurity professionals are not required to author new content, as existing scenarios can be used. They are not needed to validate that the content is comprehensive, and they do not need to build an answer key for purposes of scoring the scenario. Of course, these elements are potentially very useful; knowledge elicitation research involving cybersecurity professionals should advance on all fronts. That said, if the barriers to doing task analysis can be reduced, especially given the short supply of cybersecurity experts, then more data could be collected.

A second challenge is diversity across environments. CTA benefits from sharing (i.e., diverse SOCs need to be represented) and from momentum (i.e., CTAs can be harder to integrate if many years pass between them, as the domain changes). One next step that is needed in CTA is a method of better understanding commonalities across cyber domains. Some key research questions are: What are the operational differences between a military cyber protection team and an enterprise organization; how does sophistication in methods vary as a function of size of the organization; how does the regulatory environment or industry (e.g., banking versus education) impact cybersecurity outcomes; and, how do other organizations match or differ from past CTA? Answering these questions will require greater participation and collaboration among researchers and professionals of diverse perspectives, hallmarks of interdisciplinary science.

As an exploratory technique, CTA invites stakeholders to share their expertise to represent how the knowledge, skills, and attitudes of people facilitate cybersecurity. CTA can be a powerful complement to quantitative research as described in the next section.

Recommendation 4: Develop and Validate Diagnostic Measures of Decision-Making and Use Them for Training and Selection

Qualitative methods, like task analysis, provide an important piece of the puzzle and support quantitative investigation. However, certain relationships, such as causal linkages and generalizations from samples to populations require quantitative investigation. The primary method is null hypothesis significance testing to generalize observations of a sample to a larger population. This is a process of deductive reasoning where a hypothesis is asserted and then evidence in support of the hypothesis is obtained. Quantitative research is essential for the development and validation of measures and includes tools for measuring reliability. It also supports the evaluation of interventions.

The major benefit of quantitative research is causal description. Randomized experiments provide the best evidence for causal relationships. While randomized experimentation is powerful, it is challenging to conduct in operational environments. One need that qualitative research can provide is operationalization, the representation of constructs of interest as variables, manipulations, and measures.

CONCLUSIONS

In this chapter, I have argued for an interdisciplinary science of cybersecurity that incorporates human-centered cybersecurity. This need is critical because of the criticality of the cybersecurity problem. Big challenges include the workforce shortage, a lack of understanding about the value of people, and a lack of knowledge of what can or cannot be automated effectively.

The research agenda presented here offers many avenues for leveraging existing frameworks from other domains, developing new specialized frameworks for cyber, and conducting additional research to generate new measures and interventions. This process requires both qualitative task analysis and quantitative methods. It is also an invitation to researchers who investigate cognition that their methods are needed to improve security in this important problem space.

REFERENCES

Creswell, J. W., & Creswell, J. D. (2018). *Research Design: Qualitative, Quantitative, and Mixed Methods Approaches*. Gaithersberg, MD: NIST.

CyberSeek. (2024). https://www.cyberseek.org/heatmap.html

Evans, M., Maglaras, L. A., He, Y., & Janicke, H. (2016). Human behaviour as an aspect of cybersecurity assurance. *Security and Communication Networks*, *9*(17), 4667–4679. https://doi.org/10.1002/sec.1657

Fruhlinger, J. (2024). What is the CIA triad? A principled framework for defining infosec policies. https://www.csoonline.com/article/568917/the-cia-triad-definition-components-and-examples.html

Gutzwiller, R. S., Hunt, S. M., & Lange, D. S. (2016). A task analysis toward characterizing cyber-cognitive situation awareness (CCSA) in cyber defense analysts. In *2016 IEEE international multi-disciplinary conference on cognitive methods in situation awareness and decision support (CogSIMA)*. IEEE.

Helmreich, R. L. (2000a). *Culture, Threat, and Error: Assessing System Safety. Safety in Aviation: The Management Commitment: Proceedings of a Conference. London: Royal Aeronautical Society.*

Helmreich, R. L. (2000b). On error management: Lessons from aviation. *BMJ*, *320*(7237), 781–785. https://doi.org/10.1136/bmj.320.7237.781

Helmreich, R. L., Merritt, A. C., & Wilhelm, J. A. (1999). The evolution of crew resource management training in commercial aviation. *International Journal of Aviation Psychology*, *9*(1), 19–32.

Holden, R. J., & Carayon, P. (2021). SEIPS 101 and seven simple SEIPS tools. *BMJ Quality & Safety*, *30*(11), 901–910. https://doi.org/10.1136/bmjqs-2020-012538

Holden, R. J., Carayon, P., Gurses, A. P., Hoonakker, P., Hundt, A. S., Ozok, A. A., & Rivera-Rodriguez, A. J. (2013). SEIPS 2.0: A human factors framework for studying and improving the work of healthcare professionals and patients. *Ergonomics*, *56*(11). https://doi.org/10.1080/00140139.2013.838643

Kanki, B., Helmreich, R. L., & Anca, J. (2010). *Crew Resource Management*. London: Elsevier.

Klein, G., & Wright, C. (2016). Macrocognition: From theory to toolbox. *Frontiers in Psychology*, *7*. https://doi.org/10.3389/fpsyg.2016.00054

Miller, J. E., Patterson, E. S., & Woods, D. D. (2006). Elicitation by critiquing as a cognitive task analysis methodology. Cognition, Technology & Work, 8(2), 90-102.

Office of the National Cyber Director (Executive Office of the President). (2024). Report on the cybersecurity posture of the United States. https://bidenwhitehouse.archives.gov/wp-content/uploads/2024/05/2024-Report-on-the-Cybersecurity-Posture-of-the-United-States.pdf

Ofte, H. J. (2024). The awareness of operators: A goal-directed task analysis in SOCs for critical infrastructure. *International Journal of Information Security*, *23*(5), 3253–3282. https://doi.org/10.1007/s10207-024-00872-6

Onwubiko, C., & Ouazzane, K. (2022). SOTER: A playbook for cybersecurity incident management. *IEEE Transactions on Engineering Management*, *69*(6), 3771–3791. https://doi.org/10.1109/TEM.2020.2979832

Pizzi, L., Goldfarb, N., & Nash, D. (2001). Promoting a culture of safety. In K. Shojania, B. Duncan, K. McDonald & R. Wachter (Eds.), *Making Healthcare Safer: A Critical Analysis of Patient Safety Practices*. Rockville, MD: Agency for Healthcare Research and Quality.

Schuster, D. (2020). Exploring cognitive processes to develop cybersecurity defender proficiency. *Cybersecurity Skills Journal: Research and Practice*, *1*(2), 40–57.

Schuster, D., Fausett, C. M., Korentsides, J. M., Patel, S. M., Lazzara, E. H., & Keebler, J. R. (2024). Exploring cognition and proficiency in cybersecurity incident response: Description of a subject-matter expert interview. In *2023 IEEE conference on cognitive and computational aspects of situation management (CogSIMA)*. IEEE. https://doi.org/10.29007/gt9r

Schuster, D., & Keebler, J. (2019). Cybersecurity in organizations: A sociotechnical systems approach. In M. Mouloua & P. Hancock (Eds.), *Human Performance in Automated and Autonomous Systems: Emerging Issues and Practical Perspectives*. Boca Raton, FL: CRC Press.

Scott, A., Cooke, I., Wong, N., Sliwinska, K., & Schuster, D. (2018). Elicitation by critiquing: Applications to computer network defense. *Proceedings of the Human Factors and Ergonomics Society Annual Meeting*. https://doi.org/10.1177/1541931218621286

Shadish, W. R., Cook, T. D., & Campbell, D. T. (2001). *Experimental and Quasi-Experimental Designs for Generalized Causal Inference*. New York: Houghton Mifflin.

Statista. (2024). Digital population worldwide. https://www.statista.com/statistics/617136/digital-population-worldwide/

Terzioğlu, M. (2024). The effects of crew resource management on flight safety culture: Corporate crew resource management (CRM 7.0). *The Aeronautical Journal, 128*(1326), 1743–1766. https://doi.org/10.1017/aer.2023.113

Tetrick, L. E., Zaccaro, S. J., Dalal, R. S., Steinke, J. A., Repchick, K. M., Hargrove, A. K., … Wang, V. (2016). *Improving Social Maturity of Cybersecurity Incident Response Teams*. Fairfax, VA: George Mason University. https://calctraining2015.weebly.com/the-handbook.html

Trent, S., Hoffman, R. R., Merritt, D., & Smith, S. (2019). Modelling the cognitive work of cyber protection teams. *The Cyber Defense Review, 4*(1), 125–136. www.jstor.org/stable/26623071

Verizon. (2024). 2024 Data breach investigations report. https://www.verizon.com/business/resources/reports/dbir/

Vicente, K. J. (1999). *Cognitive Work Analysis: Toward Safe, Productive, and Healthy Computer-Based Work*. Boca Raton, FL: Lawrence Erlbaum Associates.

Vigliarolo, B. (2024, November 19). America's drinking water systems have a hard-to-swallow cybersecurity problem. *The Register*. https://www.theregister.com/2024/11/19/us_drinking_water_systems_cybersecurity/

Wickens, C. D. (2002). *Spatial Awareness Biases* (Report No. AHFD-02-6/NASA-02-4). Moffett Field, CA: NASA Ames Research Center.

Zimmermann, V., Schöni, L., Schaltegger, T., Ambuehl, B., Knieps, M., & Ebert, N. (2024). Human-centered cybersecurity revisited: From enemies to partners. *Communications of the ACM, 67*(11), 72–81. https://doi.org/10.1145/3665665

2 Human Factors within the Cybersecurity Domain

Leonie Boskeljon-Horst, Sidney Dekker, and Jonathan Dekker

INTRODUCTION

The basic idea of human factors is fairly straightforward. "Human error" is not a cause of trouble; rather, it's an effect of trouble deeper inside your system. Human factors acknowledges that we are not the custodians of entirely safe systems. It refuses to believe that designers and developers and engineers can somehow make our systems entirely perfect, and that our only job is to then protect those systems from erratic, unreliable human operators—with procedures, warnings, labels, instructions, protocols, documentation, risk assessments, and more (Clarke, 1999). Instead, even the earliest human factors work (Fitts & Jones, 1947; Roscoe, 1997) showed how "human errors" are systematically connected to features of people's tasks, tools, and their operational and organizational environment. What people do makes sense at the time, given their goals, their knowledge of the system, their mind-set, and focus of attention—much of which gets driven by the setting in which we make these people work.

Human factors has made us realize that if we want to deal with "human error," we shouldn't target the human; we shouldn't just focus on the operator who works at the point of risk and tell them to try harder or focus more. Instead, we need to study and examine and retool the entire expertise- and error-producing system in which those humans are configured and which imposes constraints and obligations and resource pressures and goal conflicts on the work they need to accomplish every day. This system starts with decisions and trade-offs (in design, expenditure, procurement, outsourcing, etc.) that lie well outside the time and space accessible for operators to even influence. Not that the problem is solved with just such a commitment, though. Yes, with the help of, for example, cognitive work analysis (Vicente, 1999), systems can always be made more error-resistant and error-tolerant. But the sheer (and ever-increasing) complexity and dynamics of our systems, particularly in cybersecurity, will keep outpacing our exact understanding of their sources of success and failure (Wiener, 1964).

Interestingly, and crucially, human factors has given many the idea that the main task of those concerned with safety is to stop things from going wrong (Dekker, 2015). It is to prevent errors from happening (making the system error-resistant), or—where they do happen—to prevent them from having significant consequences (i.e.,

DOI: 10.1201/9781003413707-3

making the system more error-tolerant). In the 2000s, the notion of "safety culture" tried to capture the organizational contributions to safety and risk after researchers recognized and documented their importance (Turner, 1978). "Safety culture" was quickly coopted by managerial interests, and was turned into a functionalist, measurable concept that basically added up the values, attitudes, beliefs, and behaviors around safety of people inside an organization (Henriqson et al., 2014).

Surveying these things by means of a questionnaire could never amount to "culture," of course (Guldenmund, 2007; Myers et al., 2014), but survey results gave managers and others (e.g., regulators) a sense of knowledge and control over the kinds of things that went into decisions and trade-offs which, they believed, could get an organization into safety trouble (Edwards & Jabs, 2009). It turned out that none of it was predictive of disaster, however: safety culture survey results predicted the inverse of what in fact happened on multiple occasions and in different industries (Antonsen, 2009; Boskeljon-Horst et al., 2022). Adding up stated values and beliefs about safety, in other words, not only did not constitute "culture," but also might have sold a fiction of socially acceptable and "correct" safety priorities to the surveyor, researcher, regulator, auditor, or manager (Filho & Waterson, 2018).

One tantalizing question underneath it, though, kept standing. Even if the attitudes, values, beliefs, and behaviors that we measure with a safety culture survey may not predict (and in fact often inversely predict (or get totally wrong) what is going to happen), the idea that there must be *something* in the social-operational life of the organization that helps or hinders safe performance, has been kept alive. Rather than thinking about this in traditional human-factors terms of error prevention, error resistance and error tolerance, researchers have started thinking in terms of the kinds of capacities that help produce organizational reliability (Rochlin et al., 1987), robustness, and resilience (Hollnagel et al., 2006): "Researchers and practitioners are now concerned to specify the organizational preconditions that might enhance crisis management, safe performance or risk-handling in complex and hazardous situations" (Pidgeon & Leary, 2000). This typically involves senior commitment to safety (Long, 2016); continuous reflection on operational practice through monitoring, event analysis, feedback and simulation of past lessons and future possibilities; shared care and concern for hazards and solicitude over their likelihoods and possible impacts (Weick & Sutcliffe, 2007); and realistic attitudes about norms, rules, and procedures vis-à-vis operational realities (Boskeljon-Horst et al., 2023), that is, acknowledging the difference between work-as-imagined (WGA) and work-as-done (WAD) (Hollnagel, 2012).

The embrace of safety (or security) as the presence of capacities that generally make things go well (rather than measured as the absence of negative events) is now increasingly adopted in both the scientific literature and many applied communities (Hollnagel, 2014). In the remainder of this chapter, we first reflect more on this transition and what it means for what we consider "human factors" now. Human factors, in large part, becomes concerned less with preventing predictable errors, and more with the ability to sustain operations under variable (and only partially predictable or controllable) circumstances. What are the capacities—in teams, people, and processes—that make it so? How can we identify those capacities and what must the organization do to enhance them even further?

Then we turn to one capacity in particular, and that is the capacity to learn and improve through a restorative approach to things that inadvertently did go wrong. Most organizations have considerable opportunity for improvement in that area, in large part by directing accountability questions toward their own future (instead of exacting accountability in the form of blame about something that happened in the past). Instead of asking *who* was responsible for a particular negative event, the question becomes *what* is responsible for making work go well in the future. What are the (potential) impacts that need to be fixed, who's going to do that, and how will the organization know that it's done (and done well?). Forward-looking accountability is connected to people's roles and responsibilities (Sharpe, 2003), and helps remove obstacles and identify the capacities that set people, and the organization, up for success (Dekker, 2023).

HUMAN FACTORS IN CYBERSECURITY

At current the human factors domain remains an underexplored and underappreciated field in the cybersecurity realm (Nobles, 2018). Human factors as a term describes the environmental and organizational factors, as well as the interactions between people, which influence the behavior at work. Within the cybersecurity domain human factors has two main focus areas; trying to understand and predict human decision-making and secondly on eliminating human errors and thereby reducing cyber risks. The focus on decision-making is also aimed at reducing errors by operators in the system as they are seen as fallible components (Nobles, 2018).

Often claims are made in literature that humans create weaknesses in security and risk management. This as a result of organizations having difficulty in comprehending and extenuating the risks associated with behavior in information security. These risks stem from a combination of increasing complexity of the information systems that people work in as well as the lack of investment in human aspects of cybersecurity. Additionally, the investments made by businesses to pursue a technological solution to solve behavior-based risks indicate their disregard for understanding human decision-making and the human interaction with the information systems (Nobles, 2018).

When outcomes are considered negative, human-based decision-making can inadvertently be defined as human error. According Rahman et al. (2021) a majority of the wide-ranging security incidents can be traced back to human errors such as mistakes or forgetfulness. The areas where humans interact with the system are often the most vulnerable points in creating a safe digital climate (Jeong et al., 2019; Rahman et al., 2021; Pollini et al., 2022). This often takes the form of employees intentionally or unintentionally failing to comply with their company's cybersecurity policy (Lilienthal, 2023).

As stated, the perception of human factors and human involvement in the cybersecurity industry is one that perceives individuals as fallible, error prone operators that are necessary for the creation of a cybersecurity organization/environment. Hence, to solve the problems that individuals pose, organizations choose to invest in technological solutions (Jeong et al., 2019), in an attempt to design the operators out of the equation instead of focusing on improving or aiding the individual. However, in order to better support individuals in cybersecurity organizations it is important to try to understand their work environment, as human actions are inseparable from the context. Nobles (2018) found that by understanding the work environment, and more specifically the work culture that employees operate in, organizations can get a better feeling for how employees engage/disengage in desired security-oriented behaviors. In order to elevate this further and maximize the learnings that can be taken from understanding the work environment, organizations should strive to create a collaborative organizational culture in which the focus lies on taking advantage of employee's intellectual assets. This requires the entire organization to participate in creating the collaborative culture, taking into consideration the broader organizational context as well as human factors elements, as this all relates to the effective functioning of cybersecurity management (Nobles, 2018).

More recently the cybersecurity field has started warming up to more modern ideas presented by safety science. As systems become more complex the reasons that failures occur also become more complex. More interactions take place within the system between various components that are tightly interconnected. The most important addition modern safety science has added to safety thinking in cybersecurity is creating an environment where there is room and interest in finding out why individuals acted in a certain way (Lilienthal, 2023). In addition, there is an interest in using the developments that have been made in accident investigations and hazards analysis and use these to improve cybersecurity systems. Cybersecurity should move away from trying to identify individuals as culprits when things go wrong, instead seeking to understand the wider system and how its operation led to a breakdown/incident (Lilienthal, 2023). Having an appreciation for individuals and a desire to understand their plight to operate as safely as possible, taking into consideration the wider system and its fallibilities, creates the backdrop to introduce some modern safety science thinking. These ideas will enable organizations to have a better understanding of what is happening and how to improve performance.

As stated in the introduction, the current interest in many industries lies within finding, defining and understanding the intangible *something* that hinders safe performance. In the cybersecurity domain this manifests itself in traditional thinking of seeking to prevent errors from occurring. When they do occur, these organizations use the term "human error" to pass the blame to individual operators instead of analyzing the wider system. Currently, cybersecurity organizations invest money into creating environments in which the occurrences of error are designed to be limited. Instead, organizations should be more concerned with distinguishing the conditions that lead to successful outcomes and create safe and desirable outcomes. To create success, organizations must understand how success is created in their organizations, this is done through understanding the gap between WAI and WAD as well as identifying and stimulating resilient capacities.

WAI and WAD are terms used to describe the difference in perspective that exists between front-line operators and their managerial counterparts. WAI describes the assumptions, rules, and policies that exist about how work should be done. While WAD refers to how something actually is done, this can be on occasions or habitual (Hollnagel, 2017). Generally, managers are focused on the WAI side; creating rules and policies which they believe will make the operation successful and safe. On paper this might well be the case. However, everyday operations offer challenges that are not described in these assumptions. When this happens, front-line workers have to adapt to the situation and find ways to keep the operation going. This is how work is then actually done. In a dynamic environment, such as cybersecurity, where events can take place quickly, the front-line operators have to react quickly to get the operation back to normal as soon as possible. In these cases, especially the reliance on operator expertise grows and WAD becomes more prominent. As an organization you may wonder what you can do to stimulate operator expertise and ensure the capacities are present to recover should such an event take place. Such capacities, and that of the wider system, can be encapsulated in the term resilience. Resilience can be described as the inherent capacity of an organization or system to stay at or recover to a dynamically stable state, this enables continued operations following an incident and/or during uninterrupted strain on resources (Hollnagel et al., 2006).

RESILIENCE

Resilient systems have four key characteristics which define their functioning under stress and capacity to recover from upsets. These four are: *monitoring*: the capacity to know what to look for, and monitor things that could seriously impact system performance in the short term, in a negative or positive way. *Responding*: being aware of what to do, having the capacity to respond to variations (expected and unexpected), disruptions, and chances in the system. *Learning:* being able to learn from events, successful and unsuccessful through analysis, communication and sharing of experiences. *Anticipating* having the capacity to be ready for future events be it as disruptions, limitations, or opportunities in the system (Hollnagel, 2018). The four characteristics of resilience are tools that organizations can use to increase the system's capacity to be resilient. Resilient systems and organizations look for ways to amplify the capacity to succeed under varying conditions (Hollnagel, 2015). According to Hollnagel (2015), resilience is multifaceted and resilience is something a system does and is not something a system has, which means it cannot be narrowed down to a single quality. Hollnagel elaborates further on the four key characteristics of resilient systems, as discussed above, as designated capacities. The *capacity to respond* requires either prepared responses and assets or have the capability to be flexible enough to make the necessary resources available as needed. There is a distinct need to distinguish between what is important and what is urgent during responses. The *capacity to monitor* requires observing both the systems own performance as what is happening in its surroundings. Monitoring refines the capacity the system has to manage short-term events, risks, and chances. The factors being monitored need to be reviewed from time to time to ensure they remain relevant. The *capacity to learn* impacts both the previous abilities as lessons learned from events can provide key details which change responses or what needs to be monitored. Learning needs to be done from when things go wrong and when they go right, both can pertain lessons about the resilience and performance of the system. Finally, the *capacity to anticipate* is associated with looking at the more distant future. Identifying events, circumstances and chances that can be advantageous or damaging to the system's continued functioning. The abilities as described by

Hollnagel create a broad scope from which organizations can work on seeking where they stand in terms of their capacity to perform resiliently.

The four concepts of resilience form the basis from which a resilient organization emerges. The desire to create resilient organizations stems from the inclinations that each part of an organization has, depending on their focus, for example, marketing, finance, or safety. These predispositions shape decision-making processes and forms within the organization, leaving the wider organization vulnerable to threats by impeding system thinking strategies which would ensure integration into the broader surroundings (Watts & Paciga, 2011). Resilient organizations go through four stages in their quest to form predispositions that lead to better understandings of their environment and the adaptive capabilities required to succeed in that environment. The first three of these four stages are *operating*; focus on reliability and technology through technical leadership, *managing;* focus on planning and organizing resources through systematic leadership, *leading;* focus on engagement, team building, and strategy through systemic leadership. The highest stage which an organization can achieve is *integrating and uniting.* This stage describes advanced organizations as those that understand that future proofing their operations necessitates them to establish and maintain capacitates to innovate, institute, and foresee challenges (Watts & Paciga, 2011). An organization of this level sees their employees as valuable assets that aid in defining the future direction of the organization. This is supported through investments in an environment in which the inventiveness of the employee can prosper. Organizations of this level work on shared decision-making and troubleshooting, confidence, eagerness to grasp opportunities for learning in a harmonious environment. Managing the various interests of departments and individuals effectively will lead to a model where power is distributed, and integration is key (Watts & Paciga, 2011). The four concepts of resilience as introduced by Hollnagel and the elaborations made by Watts and Paciga clarify the necessities for a resilient organization. By adhering to these principles' organizations can be better prepared for threats in the short and long term. Cybersecurity organizations can benefit a lot from the principles of resilience as well as a better understanding of human factors.

RESILIENCE AND HUMAN FACTORS IN CYBERSECURITY

A resilient cybersecurity organization would make use of the four characteristics of resilient systems: monitoring, responding, learning, and anticipating. Thereby creating an organization that is ready for challenges to its operational continuity. Such an organization would also be focused on maximizing the resilient potential of its employees. This requires a significant mindset change from the current manner in which cybersecurity organizations respond to incidents. Labeling an incident as human error, shows that an organization is yet to grasp the deeper systems flaws that caused the incident in the first place. The organization has to have the desire to dig deeper than surface level errors, if they really want to improve the performance of their system. In combination with this mindset, organizations must understand that there will be a difference between WAI and WAD. Finding out how work is actually done is critical, this is where employees use resilient capacitates to produce results even when they are placed under constraints. Such behavior can more commonly be described as local ingenuity, these processes are well established and shared among colleagues, but not with management (Boskeljon-Horst et al., 2022). Understanding the differences between WAI and WAD as well as understanding local ingenuity, will offer the organization a baseline understanding of its own operations. Once this has been established it is key for organizations to start working with the four key concepts of resilient systems.

Organizations such as the ones in cybersecurity find themselves in an environment of constant evolution. Innovation is rapid and threats can arise at any given opportunity. Therefore, it is critical to make use of available expertise in the organization to monitor for new threats and opportunities, as well as ensuring that the learning potential from incidents is used. Organizations should ensure that there is room to look at things going right as there is lots to learn from things going right, which may improve processes, policies, and the operation. In everyday operations, things inevitably can go wrong. How we as organizations and individuals set up to react when things go wrong can impact the eventual outcomes of future events. When things go wrong, how do we ensure we still appreciate the actions of individual operators? Do we seek to understand the gaps between WAI and WAD? And appreciate the resilient capacities shown that ensured that the outcome was not worse?

THE NEED FOR OPENNESS

As explained above, a resilient organization has the capacity to respond, monitor, learn, and anticipate. What connects these four capacities is the capacity to create openness within an organization. Openness means, in this context, that safety information is shared freely within an organization so that it can be used to enhance both safety and the effectiveness of an organization. As Ivan Pupulidy put it elegantly, "The currency of safety is information" (Safety Monkey podcast, episode 11, 2020). Responding is only possible if it is known what to respond to. Although large disturbances in a system will be easily spotted and hence known, this might be different for small incidents, where the possibility exists to hide what happened. The same can be said for monitoring, what is needed is information that shows what is important to monitor.

Openness and shared safety information become even more vital when it comes to the capacity to learn and anticipate. As the literature shows, learning from occurrences allows organizations to improve safety (Dekker, 2006; Reason, 1997; Woods et al., 2010), which is particularly key to organizations operating in high-risk environments (Catino & Patriotta, 2013). A continuous flow of safety-related information is vital to this learning (Dekker, 2002). Operators involved in occurrences possess this safety-related information, as they are the closest witnesses when these negative events unfolded. How an organization responds to an occurrence determines the quantity and quality of information that runs freely throughout the organization. It also determines how safe people feel to share their concerns (Boskeljon-Horst, 2023). The question then becomes how to create such an open climate, in which employees are willing to share the relevant information they possess?

THE CAPACITY TO BE JUST

The question on how to create openness and gaining access to relevant safety information is not a new one. In the 80s and 90s of past century a dilemma was perceived between sharing information about occurrences and possibly having to appoint responsibility and accountability if the person with the information was actually involved in the occurrence. If an occurrence such as an incident occurs, it is expected that however caused it, is being held accountable for it and possibly faces appropriate measures or consequences (Dekker, 2007c). Because operators can be punished based on the information they could reveal, after all, they are the one to blame for negative events, the open sharing of safety-related information is hindered (Boskeljon-Horst, 2023). Therefore an idea rose that if safety information is being shared, for instance in the organizational reporting system, a no-blame attitude should apply. This no-blame attitude would ensure the future reporting of occurrences or other relevant safety information (Dekker, 2007c). This no-blame response was deemed appropriate as the complexity of organizational systems and hence the differences between WAI and WAD made it rather difficult to blame operators for undesired outcomes. After all, complexity results in limitations in an organizational system and until all the limitations of a system have been engineered away, it is considered unreliable and operator behavior, therefore, is not a cause but a symptom of underlying systemic problems (Reason, 1997). The only time where the no-blame attitude does not apply is in cases of gross negligence (Reason, 1997).

This no-blame attitude in other cases than gross negligence, however, contradicts with the perceived need to hold people accountable for the smaller incidents as well. Blame and (mostly) no-blame were therefore seen as irreconcilable. Being able to hold people accountable and at the same time ensure their willingness to voluntarily report safety-related issues without being punished for it required a compromise between the blame-free and blame approaches (Dekker & Breakey, 2016) (International Civil Aviation Organization, 2018). This resulted in Reason (1997) introducing the concept of just culture to solve this tension between blame and no-blame.

According to Reason, a just culture is a clear understanding between acceptable and unacceptable behavior and, at the same time, in which people are encouraged for providing essential safety information (Reason, 1997). This creates an atmosphere of trust, where events are shared and reported (Kováčová et al., 2019) and in which there is clarity on what is and is not acceptable (Balcerzak, 2017; EUROCONTROL, 2006; Reason, 1997). The concept of just culture has been embraced by sectors such as aviation, healthcare, nuclear organizations, and the railway (Heraghty et al., 2021).

Supporting the concept of just culture, Reason devised a decision tree as an analytical tool to be used to determine if behavior is either acceptable or unacceptable. One goes through this decision tree by answering different questions. Determining whether or not behavior is considered acceptable depends on how the questions about the occurrence and the involved individual are answered. With models like this decision tree an individual can be both blamed and exonerated for whatever negative event happened. Either way, learning occurs, be it from understanding how operators get caught up in an undesired event (no-blame) or by making clear what constitutes unacceptable behavior (blame) (Boskeljon-Horst, 2023). This just culture approach can be considered a fairly quick, clear-cut process with the consequences of errors

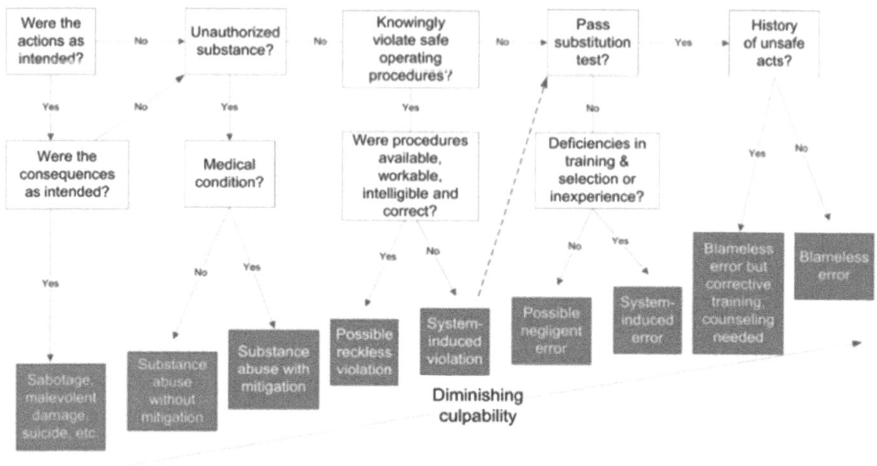

FIGURE 2.1 Traditional just culture: a decision tree for determining the culpability of unsafe acts. The answers to the questions determine the action that needs to be taken (Reason, 1997).

clear to everyone (International Civil Aviation Organization, 2018). Acceptable and non-acceptable behavior are clearly defined (Reason, 1997) and the disturbed balance, between behavior and consequences, is restored by holding someone accountable (Dekker, 2016). As long as the consequences are proportionate to the actions, a clear application of the acceptable/non-acceptable line, retribution is seen as just (Dekker, 2016). The extent to which the response to an occurrence and hence safety information is considered "just" by employees will affect the willingness of these employees to share this information (Dekker, 2016).

The just culture concept, supported by tools such as the decision tree (Figure 2.1), is supposed to ensure both the justness and forthcoming openness by, for instance, protecting employees as reasonably as practicable from disciplinary measures by providing a clear and known line between acceptable and non-acceptable behavior (Reason, 1997).

HOW JUST IS A JUST CULTURE?

As explained above, just culture models, as originally developed by Reason, are designed to determine whether an event is classified as a mistake or as negligence and that draws a clear line between acceptable and non-acceptable behavior (Marx, 1997; Reason, 1997). This has resulted in criticizing the concept of just culture. Regardless of the outcome, blame or no-blame, the decision tree judges individual behavior, leading one to question who does the judging and who decides what is acceptable or not (Dekker, 2009). Acceptable and unacceptable behavior are not fixed categories but are determined by time, context, and the person making the judgment (Alicke et al., 2008; Dekker, 2003b, 2009). When a judgment seems unjust, the information flow is hampered because people might try to hide their mistakes to avoid negative consequences (Dekker, 2007c).

In other words, if the consequences are perceived as unjust this can be counterproductive to the intended open culture (Brborović et al., 2019; Lawrenson & Braithwaite, 2018; Lipira & Gallagher, 2014; Reason, 1997). The fear of repercussions (retribution) can lead to quick confessions with minimal disclosure motivated by the thought that the sooner the act is reported, the more often it is considered brave and the consequences will not be that bad. The emphasis is on avoiding consequences instead of sharing information (Dekker & Laursen, 2007), clearly showing the impact of retribution on reporting culture.

Furthermore, punishing and learning are fundamentally at odds. Whereas punishing has to do with continually believing the system is safe, learning has to do with changing these beliefs (Dekker, 2003a). Punishing means moving away from the adverse event instead of confronting it (Dekker, 2007a), providing the illusion of control and fear instead of awareness (Dekker, 2007b). In short, the just culture model as described above leads to nothing less than a retributive just culture (Dekker, 2016).

In a retributive just culture, an individual is seen as the (possible) cause of an incident. The justice ("just") principle focuses on how the harm caused is compensated with a consequence that, as a form of punishment, is the price to be paid by involved individuals (Boskeljon-Horst et al., 2023). If the harm caused and consequences are considered equal in "weight," meaning the consequence is proportionate to the harm, the disturbed balance as a result of the incident is restored. To determine the appropriate course of action a set of questions is asked after an incident has occurred:

- Which rule was broken?
- Who did it?
- How bad is the violation?
- What penalty is involved/appropriate?
- Who gets to decide on the penalty? (Dekker, 2016)

The result is that people involved are held accountable for what they have done which, if this is perceived as unjust, has a detrimental impact on the sharing of safety information. In other words, a retributive just culture can have a negative impact on the openness needed for the resilience of a system. Furthermore, in this way, the accountability is something that is imposed upon the employee involved, the answer to the second question. When accountability is imposed it is less likely it is embraced by the employee involved.

RESTORATIVE JUSTICE AS AN ALTERNATIVE RESPONSE

To counter the negative effects of retribution, a second elaboration has emerged within the just culture domain, titled restorative (recovery-oriented) just culture. In a restorative just culture, the focus is not so much on judging behavior, whether this is right or wrong, but more on learning and recovery (Dekker, 2016; Dekker & Breakey, 2016; Goodstein & Aquino, 2010). To determine the appropriate course of action, from a restorative perspective an entire different set of questions is asked after an incident has occurred:

- Who has suffered harm?
- What do they need?

- Who is responsible to meet those needs?
- What role do the organization and environment play in recovery and learning from this situation?

Within this approach, conversations take place between all those affected by the incident, with the aim of understanding what happened, understanding the context in which the incident took place, and answering what to do next (Parkinson & Roche, 2004). Any measures and remedial actions deemed necessary flow from analysis. Possible measures and actions are widely supported because they are the result of a group effort, not reserved for a punitive superior. By involving everyone affected by an event this becomes a broadly based decision; the "offender" participates in deciding what needs to be done, including the consequences for the offender (Boskeljon-Horst et al., 2023), delivering more satisfying results for all concerned (Braithwaite, 2002). Research shows that even with major events such as sabotage or vicious crimes, the restorative approach has more satisfying results than the retributive approach for the victims, the offenders and communities (Braithwaite, 2002).

Accountability also plays a major role in restorative justice culture, but the way in which this is done differs from the retributive approach. People are still held accountable, but instead of focusing on punishment imposed by an impartial party, which in the case of severe accidents and crimes is hardly ever satisfactory for the victims (Braithwaite, 2002), restorative justice focuses on what needs to be done to repair trust and relationships. Because the focus is not on punishment and because involved employees are part of the restorative process, chances are very high they will give an account of what happened, of their role in the incident. This means that instead of accountability being imposed, accountability is embraced.

Safety within organizations is positively affected by a restorative just culture approach. For instance, trust, shared accountability and learning is increased (Turner et al., 2020), a focus on repairing and learning results in both quantitatively and qualitatively richer information to detect near misses, system solutions, and a safer workplace (Bell et al., 2011; Dekker & Breakey, 2016) and furthermore fewer suspensions and disciplinary measures, reduced absences due to illness, and increased disclosure and cost-effectiveness (Kaur et al., 2019). The impact of a restorative just culture on reporting behavior is a positive one: research shows that a restorative response results in a dialogue which allows an organization to identify all sides of an event by combining all perspectives involved (Dekker, 2023).

ENSURING A RESTORATIVE RESPONSE

The goals of a restorative response are (Dekker, 2023):

1. Fix what has been impacted by the incident

   Besides material damage, people will be damaged as well, if not physically at least emotionally. An incident can have both first victims, people directly impacted by the incident, and second victims, people involved in the occurrence of an incident. Second victims might suffer from feelings of guilt and responsibility for the incident. Both first and second victims have needs such as understanding, recognition, the ability to tell their story, and attending to these needs is part of fixing what has been impacted by the incident. Attending to these needs is necessary to put the incident behind and focus on the future.

2. Moral engagement of those involved in and impacted by the incident

   Moral engagement has to do with accepting appropriate responsibility. Within the restorative respective, this responsibility runs much wider than the responsibility of the individual operator or worker. It goes all the way up to management responsible for decisions regarding budget and tooling for instance, which may have aided the incident to occur. Moral engagement has also to do with recognizing and acknowledging the impact of the incident on people within as well as beyond the own working environment. This means that collecting the different perspectives and responding to each and every one of them. Finally, moral engagement means humanizing the people involved in the incident. Employees involved in the incident are not culprits, they are not numbers on a spreadsheet, they are people emotionally affected by the incident. Humanizing them means being able and willing to listen with honest curiosity and interest and without judgment to their account of what happened.

3. Worker reintegration into the operation

   Workers are usually trained and skilled for their job, which means that the organization has made an investment in them. Losing this investment due to a perceived unjust treatment after an incident is not only an economical waste but also a practical one. As an organization you not only have to invest again, but hiring new workers might prove a challenge. An unjustly treated worker is usually not a good ambassador for an organization and might provide negative pr. Finally, the worker involved in an incident has a unique perspective on what happened. Not reintegrating that worker means losing that perspective and hence valuable information. That in itself is reason enough to take care of a worker involved in an incident.

4. Organizational learning and enhancing capacities and resilience

   A restorative response provides an opportunity to learn because restorative just culture goes beyond the directly involved workers and their actions. It considers the role and influence of many other factors as well, a systems view, whereas a retributive just culture often does not. Because of the different perspectives included in a restorative process, the underlying systemic causes of an incident or unwanted outcome are more likely to be identified and these causes can hence be addressed and changed.

Responding in a restorative manner, hence a restorative just culture is easy in theory and can be challenging in practice, especially in hierarchical organizations which seem to have a more natural tendency toward a retributive just culture (Boskeljon-Horst et al., 2023).

To ensure a restorative response the first, vital, thing to do is resist blaming humans or "human error" after an incident (Dekker, 2023). This requires acknowledgment of the complexity of an organization, of a difference between WAI and WAD and of the fact that employees do not operate in a so-called vacuum. This means that the first step toward a restorative response is actively trying to understand why it made sense for employees to do what they did (Dekker, 2023). Postpone your own hypotheses about the onset of an incident and ask involved employees to provide their perspective.

A second aspect, precondition so you will, necessary to ensure a restorative response is a non-judgmental attitude. Upon learning why it made sense to employees to act in a certain way it may be tempting to qualify this as either right or wrong. A judgment like that automatically invokes a retributive response.

Questions to ask to find out if a retributive response, although expected, is warranted or reasonable are:

- Is the work to be done and expected output possible considering the way it is described?
- Did you as a leader have any part in or influence on the situation in which the incident could happen?
- Did context or even culture have an influence on the situation in which the incident could happen?
- Did aspects such as selection, training, rules and procedures, tools, and budget have an influence on the situation in which the incident could happen?

If you are answering one or more of these questions in an affirmative way, it means that the behavior shown by an employee is not an isolated case. It did not happen in a vacuum situation. In other words, the incident is the result of an interaction of at least two, most likely many more, aspects, meaning that only blaming a worker (retributive response) is not right. It is not just. A restorative response is therefore fitting to ensure learning and future openness within the organization.

CONCLUSION

Human factors sees human error as systematically connected to features of people's tasks, tools, and their operational and organizational environment. This means that, from this perspective, human error is an effect of trouble deeper inside your system. Dealing with "human error" requires thinking about capacities that help produce organizational reliability, robustness and resilience, the capacities that generally make things go well. In particular, the capacity to monitor, respond, learn and anticipate are key characteristics of resilient systems. What binds these four capacities is the capacity to create openness within an organization. How an organization responds to an occurrence determines the quantity and quality of information that runs freely throughout the organization. If the response is considered unfair (unjust), the information flow is hampered. Ensuring openness and providing a basis to learn requires a restorative just culture in which the focus is not on judging behavior to be acceptable or not, but on learning and recovery. Safety within organizations is positively affected by a restorative just culture approach. The impact of a restorative just culture on reporting behavior is a positive one: research shows that a restorative response results in a dialogue which allows an organization to identify all sides of an event by combining all perspectives involved. A restorative response is therefore fitting to ensure learning and future openness within the organization. Human factors then, is about realizing that "human error" is inevitable and that the "antidote" lies in an open and restorative climate which ensures the resilient capacities needed.

REFERENCES

Alicke, M. D., Buckingham, J., Zell, E., & Davis, T. (2008). Culpable Control and Counterfactual Reasoning in the Psychology of Blame. *Personality and Social Psychology Bulletin*, *34*(10), 1371–1381. https://doi.org/10.1177/0146167208321594

Antonsen, S. (2009). Safety Culture Assessment: A Mission Impossible? *Journal of Contingencies and Crisis Management*, *17*(4), 242–254. https://doi.org/10.1111/j.1468-5973.2009.00585.x

Balcerzak, T. (2017). Does Electric Car Produce Emissions? *Scientific Journal of Silesian University of Technology. Series Transport 94, 187–197.*

Bell, S. K., Delbanco, T., Anderson-Shaw, L., McDonald, T. B., & Gallagher, T. H. (2011). Accountability for Medical Error: Moving beyond Blame to Advocacy. *Chest*, *140*(2), 519–526. https://doi.org/10.1378/chest.10-2533

Boskeljon-Horst, L. (2023). *The Bogeyman Unveiled - Safety and Effectiveness within the Royal Netherlands Air Force.* Delft University of Technology, Netherlands.

Boskeljon-Horst, L., Boer, R. J. D., Sillem, S., & Dekker, S. W. A. (2022). Goal Conflicts, Classical Management and Constructivism : How Operators Get Things Done. *Safety*, *8*(2), 37.

Boskeljon-Horst, L., De Boer, R. J., Steinmetz, V., & Dekker, S. W. A. (2023). Aircrews, Rules and the Bogeyman: Mapping the Benefits and Fears of Noncompliance. *Safety MDPI*, *9*(1), 15.

Boskeljon-Horst, L., Sillem, S., & Dekker, S. W. A. (2022). 'Ladder'-Based Safety Culture Assessments Inversely Predict Safety Outcomes. *Journal of Contingencies and Crisis Management*, *January*, 1–20, 372-391.

Boskeljon-Horst, L., Snoek, A., & Baarle, E. van. (2023). Learning from the Complexities of Fostering a Restorative Just Culture in Practice within the Royal Netherlands Air Force. *Safety Science*, *161*, 106074.

Braithwaite, J. (2002). *Restorative Justice and Responsive Regulation.* Oxford University Press, Oxford, England.

Brborović, O., Brborović, H., Nola, I. A., & Milošević, M. (2019). Culture of Blame—an Ongoing Burden for Doctors and Patient Safety. *International Journal of Environmental Research and Public Health*, *16*(23). https://doi.org/10.3390/ijerph16234826

Catino, M., & Patriotta, G. (2013). Learning from Errors: Cognition, Emotions and Safety Culture in the Italian Air Force. *Organization Studies*, *34*(4), 437–467. https://doi.org/10.1177/0170840612467156

Clarke, L. (1999). *Mission Improbable: Using Fantasy Documents to Tame Disaster.* University of Chicago Press, Chicago, USA.

Dekker, S. W. A. (2002). *Punishing people or learning from failure? The choice is ours.* 1–18. http://scholar.google.com/scholar?hl=en&btnG=Search&q=intitle:Punishing+people+or+learning+from+failure?+The+choice+is+ours#0

Dekker, S. W. A. (2003a). Failure to Adapt or Adaptations that Fail: Contrasting Models on Procedures and Safety. *Applied Ergonomics*, *34*(3), 233–238. https://doi.org/10.1016/S0003-6870(03)00031-0

Dekker, S. W. A. (2003b). When Human Error Becomes a Crime. *Human Factors and Aerospace Safety*, *3*(1), 83–92.

Dekker, S. W. A. (2006). *The Field Guide to Understanding Human Error.* Ashgate Publishing Limited, Farnham, United Kingdom.

Dekker, S. W. A. (2007a). Criminalization of Medical Error: Who Draws the Line? *ANZ Journal of Surgery, 77*(10), 831–837. https://doi.org/10.1111/j.1445-2197.2007.04253.x

Dekker, S. W. A. (2007b). Discontinuity and Disaster: Gaps and the Negotiation of Culpability in Medication Delivery. *Journal of Law, Medicine and Ethics, 35*(3), 463–470. https://doi.org/10.1111/j.1748-720X.2007.00168.x

Dekker, S. W. A. (2007c). *Just Culture; Balancing Safety and Accountability.* Ashgate Publishing Limited., Farnham, United Kingdom

Dekker, S. W. A. (2009). Just Culture: Who Gets to Draw the Line? *Cognition, Technology and Work, 11*(3), 177–185. https://doi.org/10.1007/s10111-008-0110-7

Dekker, S. W. A. (2015). *Safety Differently. Human Factors for a New Era.* Apple Academic Press Inc, United States.

Dekker, S. W. A. (2016). *Just Culture: Restoring Trust and Accountability.* CRC Press Taylor & Francis Group, Boca Raton, Florida, USA.

Dekker, S. W. A. (2023). *Stop Blaming. Create a restorative just culture.*

Dekker, S. W. A., & Breakey, H. (2016). Just Culture:" Improving Safety by Achieving Substantive, Procedural and Restorative Justice. *Safety Science, 85*, 187–193. https://doi.org/10.1016/j.ssci.2016.01.018

Dekker, S. W. A., & Laursen, T. (2007). From Punitive Action to Confidential Reporting. *Patient Safety & Quality Healthcare, 45*(19), 1–8. http://cla.sagepub.com/cgi/doi/10.1177/0069477007310054

Edwards, M., & Jabs, L. B. (2009). When Safety Culture Backfires: Unintended Consequences of Half Shared Governance in a High Tech Workplace. *The Social Science Journal, 46*, 707–723.

EUROCONTROL. (2006). Establishment of "Just Culture" Principles in ATM Safety Data Reporting and Assessment - EAM2/GUI 6. In *Establishment of "Just Culture" Principles in ATM Safety Data Reporting and Assessment - EAM2/GUI 6.* (Issue March 2006). https://skybrary.aero/sites/default/files/bookshelf/235.pdf

Filho, A. P. G., & Waterson, P. (2018). Maturity Models and Safety Culture: A Critical Review. *Safety Science, 105*(June), 192–211. https://doi.org/10.1016/j.ssci.2018.02.017

Fitts, P. M., & Jones, R. E. (1947). *Analysis of factors contributing to 460 "pilot error" experiences in operating aircraft controls (TSEAA94-12).* Aero Medical Laboratory, Air Material Command, Wright-Patterson Air Force Base.

Goodstein, J., & Aquino, K. (2010). And Restorative Justice for All: Redemption, Forgiveness, and Reintegration in Organizations. *Journal of Organizational Behavior, 31*(4), 624–628. https://doi.org/10.1002/job.632

Guldenmund, F. W. (2007). The Use of Questionnaires in Safety Culture Research - an Evaluation. *Safety Science, 45*(6), 723–743. https://doi.org/10.1016/j.ssci.2007.04.006

Henriqson, É, Schuler, B., van Winsen, R., & Dekker, S. W. A. (2014). The Constitution and Effects of Safety Culture as an Object in the Discourse of Accident Prevention: A Foucauldian Approach. *Safety Science, 70*, 465–476. https://doi.org/10.1016/j.ssci.2014.07.004

Heraghty, D., Dekker, S. W. A., & Rae, A. (2021). Modifying an Accident Process and Its Justice System – From Single Narratives and Retribution to Multiple Stories and Restoration. *Safety Science, 139*(April 2020), 105248. https://doi.org/10.1016/j.ssci.2021.105248

Hollnagel, E. (2012). Resilience engineering and the systemic view of safety at work: why work-as-done is not the same as work-as-imagined. *Paper Presented at the Gestaltung Nachhaltiger Arbeitssysteme, 58. Kongress Der Gesellschaft Für Arbeitswissenschaft*, pp. 19–24.

Hollnagel, E. (2014). *Safety-I and Safety-II The Past and Future of Safety Management.* Taylor & Francis Group.

Hollnagel, E. (2015). RAG – Resilience Analysis Grid. *Resilience Engineering in Practice: A Guidebook*, 275–295.

Hollnagel, E. (2017). Can We Ever Imagine How Work Is Done? *Hindsight*, *25*, 10–13.

Hollnagel, E. (2018). *Safety-II in Practice Developing the Resilience Potentials*. Routledge Taylor & Francis Group.

Hollnagel, E., Woods, D. D., & Leveson, N. G. (2006). *Resilience Engineering: Concepts and Precepts*. Ashgate Publishing Co.

International Civil Aviation Organization. (2018). *Fostering Just Culture in Operators And Service Providers*. https://www2023.icao.int/APAC/Meetings/2018%20APRAST12/APRAST12%20WP-11%20AI_5%20-%20[NokScoot_AEROTHAI_CANSO]%20Just%20Culture.pdf - [NokScoot_AEROTHAI_CANSO] Just Culture.pdf#search=just culture

Jeong, J., Mihelcic, J., Oliver, G., & Rudolph, C. (2019). Towards an Improved Understanding of Human Factors in Cybersecurity. *IEEE 5th International Conference on Collaboration and Internet Computing (COC)*, 338–345.

Kaur, M., De Boer, R. J., Oates, A., Rafferty, J., & Dekker, S. W. A. (2019). Restorative Just Culture: a Study of the Practical and Economic Effects of Implementing Restorative Justice in an NHS Trust. *MATEC Web of Conferences*, *273*(April 2018), 01007. https://doi.org/10.1051/matecconf/201927301007

Kováčová, M., Licu, A., & Bálint, J. (2019). Just Culture - Eleven Steps Implementation Methodology for Organisations in Civil Aviation - "jC 11". *Transportation Research Procedia*, *43*, 104–112. https://doi.org/10.1016/j.trpro.2019.12.024

Lawrenson, A. J., & Braithwaite, G. R. (2018). Regulation or Criminalisation: What Determines Legal Standards of Safety Culture in Commercial Aviation? *Safety Science*, *102*(July 2016), 251–262. https://doi.org/10.1016/j.ssci.2017.09.024

Lilienthal, M.. (2023). Integrating Safety into Cybersecurity Test and Evaluation. *The Journal of Test & Evaluation*, *44*(2).

Lipira, L. E., & Gallagher, T. H. (2014). Disclosure of Adverse Events and Errors in Surgical Care: Challenges and Strategies for Improvement. *World Journal of Surgery*, *38*(7), 1614–1621. https://doi.org/10.1007/s00268-014-2564-5

Long, R. (2016). The Social Psychology of Risk, Safety and Leadership Maturity. *The AusImm Bulletin*, *12*(5), 26–28.

Marx, D. (1997). Discipline: The role of Rule Violations. *Ground Effects: Reporting Maintenance and Groundcrew Error Reduction Efforts*, *2*, 1–3. http://www.system-safety.com/articles/GroundEffects/Volume 2 Issue 4.pdf

Myers, D. J., Nyce, J. M., & Dekker, S. W. A. (2014). Setting Culture Apart: Distinguishing Culture from Behavior and Social Structure in Safety and Injury Research. *Accident Analysis and Prevention*, *68*, 25–29. https://doi.org/10.1016/j.aap.2013.12.010

Nobles, C. (2018). Botching Human Factors in Cybersecurity in Business Organizations. *HOLISTICA – Journal of Business and Public Administration*, *9*(3), 71–88. https://doi.org/10.2478/hjbpa-2018-0024

Parkinson, J., & Roche, D. (2004). Restorative Justice: Deliberative Democracy in Action? *Australian Journal of Political Science*, *39*(3), 505–518. https://doi.org/10.1080/1036114042000295101

Pidgeon, N. F., & Leary, M. (2000). Man-made Disaster: Why Technology and Organizations (Sometimes) Fail. *Safety Science*, *34*(1–3), 15–30.

Pollini, A., Callari, T. C., Tedeschi, A., Ruscio, D., Save, L., Chiarugi, F., & Guerri, D. (2022). Leveraging Human Factors in Cybersecurity: An Integrated Methodological Approach. *Cognition, Technology and Work*, *24*(2), 371–390. https://doi.org/10.1007/s10111-021-00683-y

Rahman, T., Rohan, R., Pal, D., & Kanthamanon, P. (2021). Human Factors in Cybersecurity: A Scoping Review. *ACM International Conference Proceeding Series*. https://doi.org/10.1145/3468784.3468789

Reason, J. (1997). *Managing the Risks of Organizational Accidents.* Ashgate, Farnham, United Kingdom.
Rochlin, G. I., La Porte, T. R., & Roberts, K. H. (1987). The Self-Designing High-Reliability Organization : Aircraft Carrier Flight Operations at Sea. *Naval War College Review, 40*(4), 75–90. https://digital-commons.usnwc.edu/cgi/viewcontent.cgi?referer=&httpsredir=1&article=4373&context=nwc-review
Roscoe, S. N. (1997). The Adolescence of Engineering Psychology. In S. M. Casey (Ed.), *Volume 1, Human Factors History Monograph Series* (pp. 1–9). Human Factors and Ergonomics Society, Santa Monica, CA, USA.
Sharpe, V. A. (2003). Promoting Patient Safety: An Ethical Basis for Policy Deliberation. *Hastings Center Report, 33*(5), S2–19.
Turner, B. A. (1978). *Man-Made Disasters.* Wykeham Publications, London.
Turner, K., Stapelberg, N. J. C., Sveticic, J., & Dekker, S. W. A. (2020). Inconvenient Truths in Suicide Prevention: Why a Restorative Just Culture Should Be Implemented Alongside a Zero Suicide Framework. *Australian and New Zealand Journal of Psychiatry, 54*(6), 571–581. https://doi.org/10.1177/0004867420918659
Vicente, K. J. (1999). *Cognitive Work Analysis: Toward Safe, Productive, and Healthy Computer-Based Work.* Lawrence Erlbaum Associates, Mahwah, NJ, USA.
Watts, G., & Paciga, J. J. (2011). Conscious Adaptation: Building Resilient Organizations. *AAAI Fall Symposium - Technical Report, FS-11-03*, 158–167.
Weick, K. E., & Sutcliffe, K. M. (2007). *Managing the Unexpected: Resilient Performance in an Age of Uncertainty* (2nd ed.). Jossey-Bass, Hoboken, NJ, USA.
Wiener, N. (1964). *God and Golem.* MIT Press, Cambridge, MA, USA.
Woods, D. D., Dekker, S. W. A., Cook, R., Johannesen, L., & Sarter, N. (2010). *Behind Human Error.* Ashgate Publishing Limited, Farnham, United Kingdom.

3 Cybersecurity Incident Response Teams and Multiteam Systems

An Input-Process-Output Framework

Stephen J. Zaccaro, Hannah C. Luckes, and Joanna Zakzewski

Cybersecurity incident response (CSIR) efforts sometimes stumble, or even fail, not because of the quality of detection and ameliorative technology or of missing knowledge and skills of individual analysts, but rather because such analysts fail to work effectively as a team. They are missing critical collaboration and communication processes and skills, especially with key stakeholders. Indeed, President Obama (2009) noted: "when it comes to cybersecurity, federal agencies have overlapping missions and don't coordinate and communicate nearly as well as they should – with each other or with the private sector." More than a decade later, Richberg (2022) noted that:

> "Cybersecurity is a team sport" is a well-worn cliché…[however], many organizations do not take this concept far enough. Given the expanding attack surface every organization faces today, the cyber skills talent gap and dynamic threat landscape, there is a growing urgency to maximize the value of teamwork as much as possible to secure organizations and important data.

Case studies of CSIR failures point to the critical role of poor teamwork in these outcomes. For example, Equifax suffered a cybersecurity breach on its Automated Consumer Interview System on May 13, 2017, but it was not detected until July 29, 2017. This breach affected 148 million consumers and, after being announced, caused an almost $6 billion loss to the company's market value (Kabanov & Madnick, 2020). While multiple factors contributed to the breach, a case analysis conducted by Kabanov and Madnick noted several key errors in team processes that delayed detection and communication of the breach, exacerbating its effects on consumers. For example, although the SSL certificates had expired in the previous year, the information security (IS) team was not aware that "the system was incapable of identifying and preventing intrusions and continued relying on its protective capability against attacks" (Kabanov & Madnick, 2020, p. 11). Zaccaro et al.'s (2016; see

DOI: 10.1201/9781003413707-4

also Tetrick et al., 2016) taxonomy of CSIR performance included the identification, testing, and communication of system vulnerabilities as a critical proactive team process. Likewise, although the IS team at Equifax did receive an alert notification from US-CERT of potential vulnerabilities in the software systems they were using, members of the team communicated the alert poorly to others in the company such that necessary urgency was not conveyed sufficiently, and relevant personnel were excluded from communications sent out by the IS team (see Kabanov & Madnick, 2020, p. 12). These failures reflect poor performance of critical reactive incident response team processes around information exchange and coordination (Tetrick et al., 2016; Zaccaro et al., 2016).

Case studies also show that effective attention to team processes does result in better cybersecurity performance. For example, Ahmad et al. (2021) described a case analysis of how an organization effectively acquires and disseminates incident response situation awareness that in turn fosters effective team resolution of a cyber threat. They defined such awareness as "perception of incident-related elements within the organizational environment over the course of the incident, comprehension of their meaning within the context of the organization's cybersecurity mission and objectives, and the projection of their status in the near future" (p. 4). In their analysis, Ahmad et al. show how knowledge sharing from these situation awareness processes by teams involved in incident response facilitates amelioration and resolution of cyber threats.

Cuevas et al. (2021) also provided evidence for how effective teamwork drives successful CSIR from interviews with members of the top five teams (out of 18,201 teams) in a computer security Capture-the-Flag competition. The members of these teams were geographically dispersed, exacerbating challenges for coordination and collaboration. Cuevas et al. (2021) found that the top teams in this competition developed effective strategies for fostering efficient work distribution, defining member roles, and using technology to facilitate task coordination.

These case studies, of both failed and successful incident response, indicate the important role team attributes and processes have in successful cybersecurity operations. Thus, understanding and enhancing the elements of effective teamwork should be a key focus in improving the overall posture of cybersecurity in organizations. This chapter describes these elements in terms of team inputs and processes that contribute to team outcomes.

However, our analysis will not only focus on individual teams. CSIR often entails the close collaboration of multiple teams. Such team networks have been defined as multiteam systems (MTSs) (Mathieu et al., 2001; see also Chen et al., 2014; Dalal et al., 2022; Tetrick et al., 2016). In such systems, failures and poor performance may occur not because of poor within-team processes, but because of inadequate coordination *between* teams. Indeed, the analysis by Kabanov and Madnick (2020) of the Equifax security breach showed that poor performance was a result of the IS and IT teams not adequately sharing information about the cyber-attack and not developing shared situation awareness of system vulnerabilities. Likewise, the analysis by Cuevas et al. (2021) demonstrated how success resulted from effective and timely sharing of information across different teams and across phases of developing situation awareness and generating responses to attacks. CSIR is typically a multiteam endeavor, particularly in the context of severe or urgent threats (Tetrick et al., 2016). Thus, creating effective incident response entails focusing on the elements of both teams and MTSs (Chen et al., 2014; Tetrick et al., 2016).

THE ELEMENTS OF CSIR TEAMS AND MTSs

There may be a tendency to consider MTSs as large, segmented teams and therefore what we know about teams should presumably apply isomorphically to MTSs. However, MTSs have features that would not pertain to teams, making them a distinct collective form (Zaccaro et al., 2012). In this section, we define both teams and MTSs and their key components.

WHAT IS A TEAM?

Kozlowski and Ilgen (2006) defined a team as (p. 79):

> (a) two or more individuals who (b) socially interact (face-to-face or, increasingly, virtually); (c) possess one or more common goals; (d) are brought together to perform organizationally relevant tasks; (e) exhibit interdependencies with respect to workflow, goals, and outcomes; (f) have different roles and responsibilities; and (g) are together embedded in an encompassing organizational system, with boundaries and linkages to the broader system context and task environment.

This definition includes several attributes that are important in thinking about cybersecurity teams. First, teams can be small enough to contain only two members. This definition does not indicate an upper limit on team size, but as teams get larger, they are more likely to be composed of sub-teams, and the degree to which such sub-teams have their own goals, the larger team may actually reflect an MTS. That will mean the key dynamics that predict effectiveness will be different, a point we return to later.

A second key point in this definition is that members typically have tasks that require interdependent action, which in turn drives the need for interaction. These interactions can foster emergent states such as shared knowledge and awareness, team trust, cohesion, and commitment to the team that are important for team effectiveness (Marks et al., 2001). Third, teams form to accomplish goals that are relevant to the embedding organization, and that in turn reflect a set of collective tasks. Members have specific roles in the accomplishment of these tasks and role enactment occurs interdependently across members. Finally, teams are embedded within organizations and, as such, their design, structure, and processes are affected by the context of higher-level dynamics.

Current definitions of cybersecurity incident response teams (CSIRTs) reflect some, but not all of the attributes specified by Kozlowski and Ilgen (2006). For example, the Software Engineering Institute (SEI) defined a computer incident response team as a "group of individuals usually consisting of Security Analysts organized to develop, recommend, and coordinate immediate mitigation actions for containment, eradication, and recovery resulting from computer security incidents" (SEI, 2014,

p. 4). This definition indicates the purpose or general goal of the team (the "containment, eradication, and recovery" of/from cyber incidents), and the collaborative processes that achieve this purpose. Interdependence of action is implied by the element of "coordination" in the definition. However, the degree of interdependence in team member actions will likely vary as a function of threat severity of a cyber event (Tetrick et al., 2016). Likewise, the embedding of CSIRTs in organizations plays a large role in establishing the context for CSIRT actions through budget and resource allocation decisions, and higher-level policy prescriptions (Klimoski & Murray, 2016; Shaikh & Siponen, 2023). Finally, the distinctions made between security operations centers (SOCs) and CSIRTs illustrate issues with team size. SOCs are typically responsible for incident detection and escalation to CSIRTs for resolution; however, in some organizations these different activities are handled by single teams (Dalal et al., 2022). When both the structures of incident response structures and the nature of threats confronted by organizations become more complex, these functions may be split among different teams (Dalal et al., 2022), possibly forming an MTS. A key point, here, is that the structures, policies, and processes of CSIRTs need to align with the incident response needs of the organization.

What Is an MTS?

Mathieu et al. (2001) defined MTSs as:

> two or more teams that interface directly and interdependently in response to environmental contingencies toward the accomplishment of collective goals. MTS boundaries are defined by virtue of the fact that all teams within the system, while pursuing different proximal goals, share at least on common distal goal, and in doing so exhibit input, process and outcome interdependence with at least one other team in the system.
>
> **(Mathieu et al., 2001, p. 290)**

This definition expresses several key features that would pertain to cybersecurity MTSs. We cover each of these features below, with implications for cybersecurity managers.

MTS Size

MTSs are composed of two or more teams. When this feature is put together with the size feature in the earlier definition of teams, it means that MTSs can be as small as four persons arraigned in two teams. This notion is counter to what others maintain about MTSs. For example, Davison et al. (2012, p. 809) argued that "large size and modular organization by specialization are fundamental attributes of multiteam systems that make an enormous difference in the nature of coordinated action."

While we wholeheartedly agree that size does make "an enormous difference" in terms of complexity of coordination, we suggest that, regarding MTSs, what is particularly critical, is the degree to which each component team perceives itself as a distinct entity, or the degree of "entitativity" that exists between the teams (Campbell, 1958; DeChurch & Zaccaro, 2013). Entitativity reflects "factors that distinguish a purely nominal collection of people from those that are perceived as – and function as – a group" (DeChurch & Zaccaro, 2013, p. 14). According to Campbell (1958) these factors are common fate, similarity, proximity, and boundary reinforcement. While this concept is used to distinguish groups and teams from non-purposive or random collections of individuals, it also applies to the delineation of teams within an MTS. Without clear entitativity among such teams, the MTS would more closely reflect a large team rather than a network of different but closely interacting teams (DeChurch & Zaccaro, 2013).

We can illustrate this concept with an example experienced by the first author as part of a larger study of CSIR MTSs (Tetrick et al., 2016). This study entailed a series of interviews with managers of such MTSs. One MTS was composed of two component teams, with each component team having two or three persons and a supervisor. When pressed as to why they were not a larger single team, the participants argued that each team had its own function, supervisor, and separate physical space, although they noted that each team collaborated closely. The separate functions of each team indicate they likely had separate work processes and distinct performance outcomes. This reflects the common fate element of entitativity and suggests that each team was separated from the other on this element. The distinction in functions suggests a greater similarity (a second element of entitativity) within teams around certain knowledge and functional expertise than between teams. Separate physical spaces for each team reflect the proximity element of entitativity. Finally, each component team having a different supervisor suggests boundary management, as interactions between teams may flow more regularly through such supervisors, and they in turn enforce a degree of boundary maintenance that upholds the perception of distinct teams within an MTS, even one as small as six-eight total members.

We do not belabor this point to argue against the notion that size is important in denoting MTSs – size does increase complexity as collectives grow; moreover, larger unit sizes increase the likelihood that they will evolve into MTSs. However, CSIR managers of smaller MTSs will often misperceive the collective as a larger team, not noting the critical distinctions among the teams, and therefore may not manage effectively the critical dynamics that need to occur between the distinct teams.

MTS Goal Structures and Component Team Interdependence

The aforementioned definition of MTSs by Mathieu et al. (2001) defines two types of goals in such collectives. All teams in MTSs are working toward a common distal goal; each component team has a contribution to make toward this goal. However, each team has its own proximal goal along the path to distal goal accomplishment. Figure 3.1 illustrates a goal hierarchy for a CSIR MTS (Zaccaro et al., 2016; note, not all potential component teams are represented in this illustration). The distal or overall goal is to restore and maintain the safety of an entity's cybersecurity infrastructure. Proximal goals that lead to distal goal accomplishment include (a) detecting malicious activity; (b) determining the nature of the threat; (c) eradicating a threat from a system; and (d) sharing details about the threat and its mitigation. Each

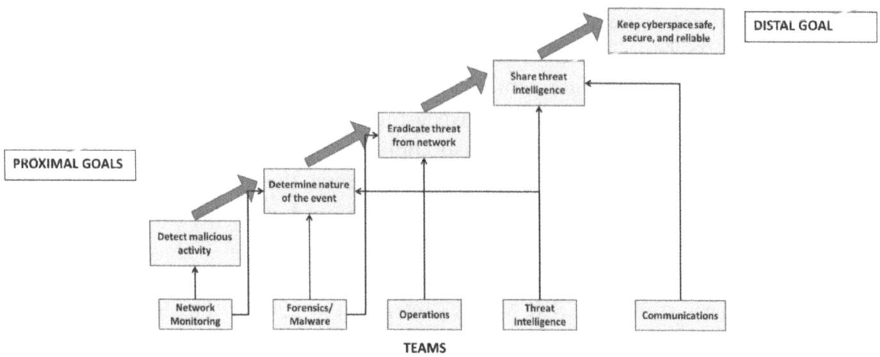

FIGURE 3.1 A cybersecurity incident response MTS goal hierarchy.

Adapted and used with permission of Taylor and Francis, from Zaccaro, et al. (2016), A Comprehensive Multilevel Taxonomy of Cybersecurity Incident Response Performance. In S. J. Zaccaro, R. D. Dalal, L. E. Tetrick, & J. A. Steinke (Eds), *The Psychosocial Dynamics of Cyber Security*; permission conveyed through Copyright Clearance Center, Inc.)

proximal goal is accomplished by one or more of the component teams, but the distal goal requires the collaborative efforts of the five teams in the MTS.

Perhaps the key element that distinguishes MTSs from other collections of teams in most organizations is the high degree of interdependence and necessary collaboration among the teams in the MTS (Zaccaro et al., 2012). All teams in an MTS work closely with at least one other team to accomplish MTS goals. Mathieu et al. (2001) specified several forms of interdependence among such teams. The first, input interdependence, refers to the integration of resources component teams bring to the MTS, including expertise, technology, financial resources, and information. Second, output interdependence, occurs when benefits and outcomes that can accrue to a component team depend upon the actions and goal accomplishment by other component teams.

Process interdependence is the third form of component team integration specified by Mathieu et al. (2001), and it is perhaps the form that most distinguishes MTSs from other organizational collectives (Luciano et al., 2018). This form of interdependence refers to the intensity of interactions and collaborations that need to occur among component teams to accomplish MTS proximal and distal goals. Mathieu et al. described three levels of process interdependence that typically occurs in MTSs (see also Tesluk et al., 1997). Sequential interdependence occurs when one component team completes a task and then passes that task along to other component teams for their contributions. For example, in CSIR MTSs, a watch or monitoring team may detect a threat and, if that threat reaches a particular threshold beyond routine levels, they may pass it along to an analysis team (Dalal et al., 2016). Reciprocal interdependence refers to the reiterative and cyclical exchange of tasks among teams. Thus, a cybersecurity engineering team that is developing threat detection software may pass along prototypes of the software to a watch team for testing. Results are then returned to the engineering team for analysis and software revision or upgrading. The third form, intensive interdependence, occurs when "the actions of component teams…transpire in simultaneous (or rapidly sequential and reciprocal) collaboration" (Zaccaro et al., 2012, p. 10). For example, after determining the nature and seriousness of a cyber threat, representatives from multiple teams, such as malware analysis, threat intelligence, and communications, may get together to determine how best to eradicate a threat from the system. Such intensive interdependence is more likely in the context of more severe and fast-moving threats (Tetrick et al., 2016).

MTS Interdependence Variability and Dynamism

The definition by Mathieu et al. (2001) indicates that each component team interacts with at least one other team in the MTS. As shown in Figure 3.1, not all teams interact with all other component teams; they collaborate directly only with those teams related to their proximal goals. This suggests that interdependence patterns among the component teams will not be the same across the MTS. Using information from interviews with managers of 28 CSIR MTS Tetrick et al. (2016) constructed a prototype of interaction patterns in such collectives, These MTSs typically contained seven component teams. Tetrick et al. found that the highest degree of interdependence was between the engineering team and the watch team, where the former develops the software to be used by the latter. The next highest levels were seen in the interactions between the threat intelligence and forensic teams, respectively, and multiple other component teams. The watch team interacted sequentially (i.e., lower level of interdependence) with the incident response team, as did the threat intelligence and forensics with the communications team.

A critical point identified by Tetrick et al. (2016) from their interviews is that these interaction patterns changed – that is, increased – as event severity increased. For example, they noted that, in high-impact cybersecurity breaches, the incident response team may exhibit a higher interaction tempo with the malware analysis, forensics, and threat intelligence teams; all these teams may also increase the intensity of their collaborations with the communications teams. Such dynamism in the interactions among component teams can occur for reasons other than the nature of the event. These include changes in the primary task of the MTS that shifts component teams' proximal goals (e.g., moving from immediate incident response to evaluating and changing software systems), changes in overall organizational policies around cybersecurity, shifts in the nature and novelty of threats that need to be addressed, and changes in external stakeholders that need to be considered in cybersecurity MTS operations (Luciano et al., 2018). These are factors that can also change the overall composition of the MTS in terms of necessary component teams that need to collaborate around particular distal goals.

These qualities of incident response MTSs pose several challenges for their managers. First, they need to be aware of and understand the structural arrangements of the teams under their command. Many managers perceive their MTSs are simply large teams without being fully cognizant of the functional differentiations among the teams, their integrated but separate proximal goals, and the requisite patterns of typical collaborations among them. Managers of such systems need to adjust their approaches to fit the contingencies, goal-related needs, and contexts of each component team, as well as manage the MTS as a whole. Thus, they need clear knowledge of their MTS structural arrangements and how the different teams collaborate to reach proximal and distal goals.

A second challenge for CSIR MTS managers is that they need to understand when and how these structural arrangements and collaboration patterns change in response to different events and goals. A useful exercise for such managers to know their MTS structural arrangements and how they might change is MTS mapping (Tetrick et al., 2016). To map their cyber-MTS, managers (a) list the component teams working together on a distal, common goal, (b) rate the typical interaction or collaboration levels occurring between each component team with each other team, and (c) rate how these levels change in the context of more severe threats, or other key MTS events. Tetrick et al. (2016) provide an example of such a mapping tool (see Addendum 2.1 on pages 60–61 in their manual).

A third key challenge for CSIR managers is to manage at multiple levels. They need to foster the interaction processes and emergent states (e.g., trust, cohesion) not only within each component team in their system, but also among the multiple component teams. They also need to manage the integration of the MTS as a whole, including facilitating the inputs into team and MTS processes, as well as their outcomes. In the next section of this chapter, we describe in more detail these multilevel input, processes, and outcomes in CSIRTs and MTSs.

INPUT-PROCESS-OUTPUT MODELS OF CSIRTs AND MTSs

Input-process-output frameworks (McGrath, 1964) are the dominant conceptual models in research on groups and teams. According to Hackman and Morris (1975; see also Mathieu et al., 2017), inputs refer to member attributes, team properties, task parameters, and environmental factors that influence how team members engage in task accomplishment. Such engagement in turn defines the processes of the team, or the task- and team-related interactions and activities of team members. Outputs refer to a range of consequences resulting from these activities. In incident response MTSs, these inputs, processes, and outcomes can be defined at both the level of each component team and of the overall MTS. These are illustrated in Figure 3.2 and described below.

CSIRT Inputs

Team members' knowledge, skills, abilities, and other personal attributes (KSAOs) represent key determinants of team processes and outcomes. Several studies and reviews have focused specifically on the KSAOs needed for CSIRTs (Dalal et al., 2022; Jose et al., 2016; Tetrick et al., 2016). Tetrick et al. (2016) completed a job analysis that included 46 focus group interviews and 88 surveys of managers and employees from different types of CSIRTs and MTSs. They identified three sets of KSAOs considered important for CSIR: (1) cognitive attributes, such as skills in learning, problem solving, reasoning, strategic analysis, and decision making; (2) social and team-oriented skills, such as collaboration, communication, perspective

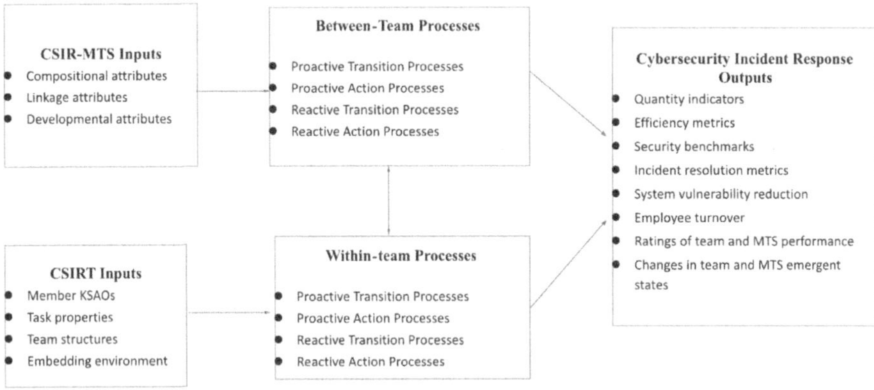

FIGURE 3.2 Input-process-output model of cybersecurity incident response team and multiteam system performance.

taking, and conflict management, and (3) character qualities, including willingness to work hard, stress management, proactivity, and persistence (see Tetrick et al., pp. 272–273, for the full list and wording of all KSAOs). Their findings suggest that team member composition consisting of these KSAOs are critical antecedents to team processes.

Dalal et al. (2022) identified two other member attributes important for CSIR performance. The first, curiosity, defined as "the desire for knowledge that motivates individuals to learn new ideas, eliminate information gaps, and solve intellectual problems" (Litman, 2008, p. 1586). This attribute is critical for CSIR because many of the cyber threat-related problems in such knowledge work are novel and complex, requiring cognitive persistence that is a quality of curiosity. Threat hunting, a more proactive incident response practice where analysts seek out and identify potential cyber threats (Milani et al., 2024), also requires high levels of individual curiosity. Such cyber threat hunters need to be attuned to anomalies that may portend threats, skill in forming hypotheses and making sense of the threats, and a high determination to find their causes and solutions (Maxam & Davis, 2024), all processes engendered by high curiosity (Kam et al., 2022).

Dalal et al. (2022) identified resilience as a second individual difference important for CSIR. They defined it as "the capacity to maintain high performance in the face of contextual shocks, or to quickly recover previous performance levels following setbacks from contextual shocks" (p. 18). Given the challenging nature of CSIR, when events can increase uncertainty, ambiguity, and proneness to severe consequences for stakeholders, members need to have the personal resources to persist through such events. Dalal et al. defined this capacity as an individual difference. However, other scholars have also defined it as a capacity of teams (Alliger et al., 2015; Stoverink et al., 2020; Zaccaro et al., 2022), MTSs (Campbell, 2024), and organizations (Räetze, et al., 2022).

Task properties, team structures, and environmental contingencies represent other key inputs to team processes (Mathieu et al., 2017). Task properties include task complexity, cyber event characteristics, and technology infusion (Dalal et al., 2022; Mathieu et al., 2017). Task complexity can result from the task (a) requiring a significant amount of information and variety of skills for its completion; (b) involving multiple subtasks or components; and/or (c) changing in its performance requirements over time (Campbell, 1988; Wood, 1986). Higher task complexity can have implications for both team composition (e.g., requiring greater skill diversity) and team processes (e.g., necessitating more interdependence in task accomplishment). Such complexity can also be engendered by the nature of cyber events and incidents. Cyber incidents can vary by their frequency (Killcrece et al., 2003) and their severity (Dalal et al., 2022). Incident severity can result from several factors, including the financial consequences of the attack, the location of the attack, its degree of reputational damage, the range and extent of populations affected by the attack, and the difficulty in uncovering its root cause and solution (Checklist Incident Priority, n.d.; Dalal et al., 2022; National Cyber Incident Scoring System, n.d.). More severe incidents are likely to require a greater range of team member skills to resolve, and more collaborations among team members and with other teams (Tetrick et al., 2016).

Team structures refer to such features as the degree of skill and authority differentiation in the team, the extent of geographical and temporal dispersion among team members, and the integration of technology into team tasks (Dalal et al., 2022; Hollenbeck et al., 2012; Mathieu et al., 2017; Mathieu et al., 2019). CSIRTs that are assigned a range of tasks related to proactive and reactive incident response (Zaccaro et al., 2016) will likely require greater skill differentiation among its members; such skill dispersion means decision authority may also be decentralized, depending on the specific expertise required for responses to particular cyber threats. Many CSIRTs are also geographically and temporally dispersed (Tetrick et al., 2016). They often work on 24/7-time cycles, where different teams are engaged at different shifts (temporal dispersion). They also may have different members and different teams located in different geographic locations, including in different countries (geographic dispersion; Tetrick et al., 2016). Both of these aspects of team structures can change the importance of certain team processes, such as interpersonal and transition processes (Marks et al., 2001), and emergent states for team effectiveness (Breuer et al., 2016, 2020; Gilson et al., 2015).

The technology infrastructure of the team is also a critical input to team processes (Dalal et al., 2022; Lane et al., 2024). Indeed, the CSIRT's security information and event management (SIEM) system is typically the conduit for triggering collaborative task activities. According to Dalal et al., the functions of a SIEM include (p. 19):

> collating cybersecurity notifications from various security technologies into a single location (referred to as log aggregation); providing audit reports to comply with laws, industry regulations, and organizational policies; and performing automated cross-correlation of raw event logs from the network to detect potential incidents.

They note, however, that SIEMs can yield a high false-positive rate, which can result in inefficient team processes as analysts waste time pursuing them (Chickowski, 2019). Thus, the accuracy, user-friendliness, and utility of reports generated by SIEMs can have significant impacts on team processes.

Certain types of CSIR-MTSs are more likely to be external MTSs than others. For example, a coordinating CSIR-MTS "coordinates and facilitates the handling of incidents, vulnerabilities, and general information across a variety of external and internal organizations, which can include other CSIRTs, vendor organizations, security experts, and even law enforcement agencies" (Bhaskar, 2005, p. 5). Thus, such MTSs are more likely to be composed of teams from different organizations. Non-coordinating MTSs can either be contained entirely within a large organization or act as managed security service providers to other organizations (Tetrick et al., 2016). Thus, the latter can also be an external MTS. Tetrick et al. (2016) noted that different types of MTSs varied on both the number of component teams they included and the particular functions of included teams. They found that coordinating MTSs were more likely to include communications, threat intelligence and malware analysis teams than other types of MTSs. They also reported from interviews with CSIR-MTS managers that "teams in coordinating CSIRT MTSs tended to display higher levels of interactions, and these interactions involved greater numbers of the component teams working together" (p. 54), supporting the role of MTS boundary status as a key CSIR-MTS input to MTS processes.

MTS compositional attributes related to boundary status include organizational diversity and proportional membership (Zaccaro et al., 2012). Organizational diversity refers to the number of different organizations representing component teams in the MTS. Coordinating CSIR-MTSs are likely to include a greater variety of organizational entities than non-coordinating MTSs (Tetrick et al., 2016). Proportional membership reflects the percentage of teams from different companies within the MTS.

Other MTS compositional attributes that serve as inputs to CSIR-MTS processes include the degree of functional diversity, cultural diversity, and geographic dispersion among the component teams. Such diversity and dispersion can exacerbate the complexity and challenges of communications and collaboration among MTS component teams (Luciano et al., 2018; Tetrick et al., 2016). Finally, according to Zaccaro et al. (2012), MTS composition can also vary in terms of the compatibility of team proximal and MTS distal goals (goal motive structure) and the amount of time and effort required by the team to accomplish MTS goals (temporal orientation) as a consequence of their functional contributions.

The remaining categories of CSIR-MTS inputs – linkage and developmental attributes – refer, respectively, to the (a) the "linking mechanisms that connect component teams" (Zaccaro et al., 2012, p. 18), and (b) "those characteristics pertaining to [MTS] developmental dynamics and patterns" (Zaccaro et al., 2012, p. 20). Earlier, we described one of these linkage attributes as the interdependencies among the component teams. Two other linkage attributes are the hierarchical arrangement of component teams in terms of their level responsibility for distal goal accomplishment, and the corresponding distribution of power and influence of some component teams over the others in the MTS. For example, as shown in Figure 3.1, two component teams, operations and communications, typically work with other teams on a single proximal goal, while others work on two or more. Responsibility for a greater number of proximal goals can give component teams greater influence over MTS-level processes and outcomes. The remaining linkage attributes of MTSs include their normative communication structures, whether centralized or dispersed among teams, and their primary

Technology has grown in sophistication to the point where artil gence and machines can act increasingly as "members" of the tear problems and accomplishing team tasks (Andrews et al., 2023; Dalal Seeber et al., 2020). Thus, the human-machine composition of the te: a critical input to team processes. Several studies have documented of effects AI team members can have on team interactions and eme (Andrews et al., 2023; Dennis et al., 2023; Harris-Watson et al., 2(et al., 2023).

A final set of inputs into team processes are characteristics of the te: ding environment. Organizations set the policies, structures, and requ their units devoted to cybersecurity and incident response. Company therefore, have significant roles as external leaders of CSIRTs (Klimosk 2016; Morgeson, 2005). According to Klimoski and Murray, the report ships of managers responsible for cybersecurity can reflect the priority o place on this function. They described six levels of reporting relatior Manager of IT operations to Chief Information Security Officer. The the best outcomes for IS occur when the C-Suite contains executive ity for such security; such a position affords the executive greater inpu pany strategic decisions, and greater opportunities to reassess cyber and investments as cyber challenges change and evolve. Thus, company exe outsized responsibility for the resources that can contribute to CSIRT ef

CSIR-MTS Inputs

The team inputs of member KSAO composition, team structural pro environmental contingencies are also critical inputs to MTS processes, tures of each team as well as the embedding environment will influence ics across the collection of teams in the MTS. However, Zaccaro et al. (2C a typology of MTS-level characteristics that serve as key inputs to MTSs beyond the properties of individual component teams. These features we into three sets: *Compositional attributes*, *linkage attributes*, and *dev attributes*.

According to Zaccaro et al. (2012), compositional attributes refer to " demographic features of the MTS, as well as the relative characteristics nent teams" (p. 13). Such features include the number of component te MTS and the overall size of the MTS in terms of total members. Larg terms of both teams and members means more potential for coordinati and process loss (Bernerth et al., 2023). Boundary status, another con attribute, refers to the degree in which component teams come from the s nization (internal MTSs) or from different organizations (external or cross MTSs). All component teams in an internal MTS are likely to be gover norms of its embedding organization, leading to at least a partial commc tion for cross-team interactions. Component teams in external MTSs m different parenting norms with greater negative potential for interteam tions. Indeed, Campbell et al. (2022) found that coordination issues that ca failures were often greater in external than internal MTSs.

communication modalities (face-to-face, hybrid, or fully virtual). The latter is determined in part by the compositional attribute of team dispersion.

Developmental attributes as MTS inputs refer to how the MTS is formed, its developmental trajectory, and the degree of constancy and dynamism in its composition and the links among teams (Zaccaro et al., 2012). Many CSIR-MTSs are formally established by their embedding organizations (or from partnerships among multiple organizations). For example, Grobler and Bryk (2010) described the formal construction of a coordinating incident response entity, and the challenges that can arise in its formation. The nature and severity of some events may entail new component teams joining an established MTS (membership constancy) or, as we noted earlier, can change the linkages among component teams (linkage constancy).

The team and MTS inputs we have outlined set the stage for different team and MTS processes. A key implication of these inputs is that managers need to be aware of MTS composition and linkage attributes to lead their MTSs effectively (DeChurch & Marks, 2006; Murase et al., 2014). Managers can use MTS mapping tools to determine the teams in their MTSs and the linkages among them (Tetrick et al., 2016). They can also assess factors that change both composition and linkages. Once they are aware of these factors, CSIR-MTS supervisors need to manage the processes and interactions both within and across component teams in the MTS. We describe these processes in more detail in the next section.

CSIRT and MTS Processes

CSIRTs typically display an interaction dynamic that has not received significant examination in the scholarly literature on teams and groups – the action of individual analysts to determine if and when team-level processes need to be initiated (Dalal et al., 2016; Tetrick et al., 2016). Such "collaboration triggering" is defined as "the process by which an individual cybersecurity analyst determines whether, when, and how to bring in other analysts or teams of analysts to mitigate an incident as a team or MTS" (Dalal et al., 2022, p. 20). Typically, analysts receiving data on potential threats need to ascertain how to respond, whether by initiating routine protocols, or by escalating the threat to other higher-level analysts and/or to other teams in the incident response MTS (Dalal et al., 2016). An escalation decision requires team norms and protocols about when to escalate events, to whom such threats should be escalated, and what information should be conveyed in escalations (Dalal et al., 2016; Dorofee et al., 2007). Such norms and protocols need to be developed as part of the CSIRT's proactive planning processes (see below). Most escalations will often trigger heightened team interactions, particularly in instances of high threat severity (Tetrick et al., 2016).

Zaccaro and colleagues (Tetrick et al., 2016; Zaccaro et al., 2016) defined a set of team and MTS processes related specifically to CSIR. They used existing team processes taxonomies from both the generic team performance literature and classifications of incident response performance (e.g., Alberts et al., 2004; Cichonski et al., 2012; DeChurch et al., 2011; Fleishman & Zaccaro, 1992; Ilgen et al., 2005; Marks et al., 2001) to define particular dimensions of team and MTS incident response processes. They also completed an initial validation of this classification. This classification was updated and revised for this chapter and is shown in Table 3.1.

TABLE 3.1
Team and Multiteam System Processes in Cybersecurity Incident Response

Processes	Proactive: Team	Proactive: MTS
Transition (Planning)	*Define team mission, tasks, and services:* share information and ideas about team mission priorities; develop shared mission awareness; develop strategies and procedures to accomplish mission goals. *Identify types of threats and vulnerabilities to security infrastructure:* Exchange information among team members about emerging security trends and potential threats to infrastructure. *Determine necessary security tools, applications, and infrastructure to achieve mission and to mitigate potential and emerging threats:* engage in information and idea sharing and evaluation about team infrastructure requirements, tools, and applications; gain collective endorsement of security measures and tools; plan team implementation of security tools, applications, and infrastructure. *Design team level procedures for maintenance and the updating of security tools, applications, and infrastructure:* engage in information sharing about (a) necessary maintenance procedures and standards, and (b) operating conditions of current security tools, applications. *Plan team structures:* establish team member roles and requirements; define team norms, performance standards, and team reward structures.	*Define MTS mission, distal goals, and services:* acquire information from component teams on proximal goals, priorities, and requirements; define MTS goal structures and reach consensus among teams; determine strategies and between protocols to meet MTS goals and mission. *Identify potential threats and system vulnerabilities that require multiteam coordination:* exchange information across component teams about emerging security trends and potential threats. Determine each component team's potential contribution to addressing future likely threats. *Determine security tools, applications, and infrastructure necessary to accomplish MTS distal goals and to mitigate potential and emerging threats:* share information across teams about necessary infrastructure requirements and operating conditions of current tools. Reach consensus on the tools, applications, and infrastructure needed for each component team to contribute to MTS distal goal accomplishment. Plan MTS level implementation of tools, applications, and infrastructure. *Design MTS level procedures for maintenance and the updating of security tools, applications, and infrastructure:* share information about and integrate the maintenance procedures and requirements of different component teams. *Plan MTS structures:* identify component teams that need to work together on mitigating typical security events; define component teams' proximal goals and roles; define MTS norms, reward structures, and performance management systems.

Action (Execution)	*Formalize team mission and goals:* implement team interaction protocols to coordinate team member actions around proximal goal accomplishment. *Implement chosen security tools, applications, and infrastructure:* assign roles to members in implementation process; coordinate members' actions in implementing tools, applications, and infrastructure; engage in collective monitoring and back-up behavior during member implementation activities. *Test for potential security vulnerabilities and threats to established security infrastructure:* coordinate member actions in conducting tests of security tools, applications, and infrastructure; determine how to communicate potential threats and system weaknesses. *Implement team structural plans:* staff the team; train team members; foster norm acceptance; implement team performance standards; foster team confidence, trust, and task-based cohesion.	*Formalize MTS mission and goals:* implement between-team interaction protocols to foster team coordination around MTS distal goal accomplishment. *Implement chosen security tools, applications, and infrastructure:* assign roles to component teams in the implementation process; coordinate and integrate the implementation of component teams' security measures, tools, applications, and infrastructure. Engage in collective monitoring and back-up behavior across component teams. *Test for potential security vulnerabilities and threats to established security infrastructure:* coordinate and integrate actions of component teams in conducting tests of the security infrastructure and information systems. Reach consensus across component teams on how to communicate potential threats and weaknesses; assign roles to component teams in such communication. *Implement MTS structural plans:* select teams to compose the MTS; train component teams according to role requirements; foster norm acceptance across the component teams; foster MTS level confidence, trust, and cohesion.

(Continued)

TABLE 3.1 ***(Continued)***
Team and Multiteam System Processes in Cybersecurity Incident Response

	Reactive	
Processes	**Team**	**MTS**
Transition (Planning)	*Gather information about security incidents:* exchange information and ideas with other team members about nature of attack; reach consensus on cases to be selected for triage and further analysis. *Triage and analyze incoming incidents:* share assessment of identified threats among team members; exchange information and reach agreement with the team on how to categorize identified events; exchange information within the team on problems and potential damage caused by the event; reach consensus within the team on the causes of the event. *Develop a comprehensive incidence remediation solution:* exchange information and reach consensus about best fitting solutions; exchange information about members' expertise and experience with potential solutions; conduct team-wide simulations of potential solutions; plan solution implementation, as well as the contributing roles and actions of key team members. *Conduct team after-action reviews:* exchange information among team members on solution effectiveness and collective activities during all phases of incident handling cases; determine necessary revisions to existing team cybersecurity policies, protocols, tools, and infrastructure. Develop implementation plan for necessary revisions.	*Gather information about security incidents:* exchange information and consult with affected and relevant component teams; reach consensus among relevant component teams on cases to be selected for triage and further analysis. *Triage and analyze incoming incidents:* share assessment of identified threats; exchange information and reach agreement across relevant component teams on how to categorize identified events; exchange information within affected component teams on problems and potential damage caused by the event; reach consensus across component teams on the causes of the event. *Develop a comprehensive incidence remediation solution:* exchange information with relevant component teams and reach consensus about best fitting solutions; exchange information about component teams' functions and expertise; conduct MTS-wide simulations of potential solutions; plan solution implementation, as well as the proximal goals, contributing roles and actions of relevant component teams. *Conduct MTS after-action reviews:* exchange information among component teams on solution effectiveness and collective activities during all phases of incident handling cases; Determine necessary revisions to existing MTS cybersecurity policies, protocols, tools, and infrastructure. Develop implementation plan for necessary revisions.

Action (Execution)	*Alert others about security incidents:* coordinate team members' activities in communication of alerts. *Communicate triage assessment and incident analysis:* coordinate updating of tickets; collaborate on creation and sharing of incident report with other stakeholders. *Implement incident remediation plan:* coordinate implementation and monitoring of best-fitting mitigation solutions. *Create incident action logs:* integrate information from team members into case summary; coordinate distribution of case summary. *Implement necessary after-action revisions:* assign member roles in revision implementation; coordinate activities and implementation monitoring among members. *Manage interpersonal dynamics in the team:* monitor and facilitate (a) communication channels and protocols, (b) conflict among team members, and (c) cohesion and morale among members.	*Alert others about security incidents:* coordinate relevant component teams' activities in communication of alerts. *Communicate triage assessment and incident analysis:* coordinate across component teams on updating case files; collaborate across component teams on creation and sharing of incident reports with other stakeholders. *Implement incident remediation plan:* coordinate component teams' implementation and monitoring of best-fitting mitigation solutions. *Create incident action logs:* integrate information from component teams into case summary; coordinate distribution of case summary. *Implement necessary after-action revisions:* assign to component teams proximal goals and roles in revision implementation; coordinate activities and implementation monitoring among component teams. *Manage interpersonal dynamics across component teams:* monitor and facilitate (a) communication channels and protocols, (b) conflict among component teams, and (c) MTS cohesion and morale.

Note. Adapted and modified from Tetrick et al. (2015), Appendix A, pp. 190–204 (see also Zaccaro et al., 2016, for an earlier version).

The classification of cybersecurity team and MTS processes includes two key dimensions: (1) transition and action processes, and (2) proactive and reactive processes. The first dimension comes from Marks et al.'s (2001) taxonomy of team processes. They defined transition processes as interactions in which "teams focus primarily on evaluation and/or planning activities to guide their accomplishment of a team goal or objective" (p. 364). Marks et al. included mission analysis, goal specification, and strategy formulation and planning in the set of team transition processes. Sparks (2015) provides an example of the use of transition processes in an Air Force CSIRT.

Marks et al. (2001) defined action processes as "activities leading directly to goal accomplishment...[These activities include]: monitoring progress toward goals, systems monitoring, team monitoring and backup responses, and coordination activities" (p. 366). These are processes used in the execution of team plans formed in the transition phases of team performance episodes. The nature of CSIR as collective "knowledge work" means that many of these incident response transition and action processes will be cognitive in nature, particularly "the perennial processing of non-routine problems" (Reinhardt et al., 2011, p. 150; Zaccaro et al., 2016). Accordingly, many of the processes noted in Table 3.1 are derived from extant classifications of complex problem-solving activities (e.g., Mumford et al., 1991, 2012).

Zaccaro et al. (2016; see also Tetrick et al., 2016) augmented these two sets of processes by describing transition and action processes that are preparatory or proactive before cyber threat occurrences versus those that are reactions to such threats. They defined proactive processes as contingency planning and actions that occur in anticipation of potential and future cyber threats. Examples include defining the team's mission and goals, identifying emerging security trends and potential threats, planning the security infrastructure needed to meet such threats, executing such plans, and creating team structures. Reactive processes include activities that include planning how to respond to actual threat events and collaborative action responses to mitigate the threat. Examples of such processes include detecting, gathering, and exchanging information about security breaches and events, analyzing the consequences of such events, developing and implementing remediation strategies, conducting after-action reviews, and implementing adaptations to existing tools and infrastructure.

The classification of collaborative cyber processes specified by Zaccaro and colleagues (Tetrick et al., 2016; Zaccaro et al., 2016) delineates processes that occur not only within CSIRTs, but also between component teams in a CSIR-MTS. MTSs often fail more because of between-team process failures than because of within-team ineffectiveness (Campbell et al., 2022). Such process failures can include the lack of preparatory planning on how teams will communicate and coordinate with one another, a lack of between-team monitoring and back-up behavior during action phases, lack of recalibration activities when changing situations require adaptation, and a lack of information sharing and joint learning activities to foster greater between-team shared cognition (Campbell et al., 2022; Torres et al., 2021).

Accordingly, teams in MTSs need to engage in proactive and reactive transition and action processes both within themselves and with other teams in the CSIR-MTS with which they share proximal goals. Examples of between-team proactive processes include reaching consensus among component teams about the mission of the CSIR-MTS and the priorities of different component teams, sharing information and ideas across component teams about emerging strategic trends, and determining and implementing MTS level security infrastructure and tools to be meet address these trends and potential threats. Examples of between-team reactive processes include gathering information across relevant component teams about security incidents and threats, including an analysis of event consequences, developing a remediation plan, and coordinating its implementation across appropriate component teams, conducting multiteam after-action reviews, and implementing adaptations across the MTS to existing tools and infrastructure.

These team and MTS proactive and reactive transition and action processes are elucidated in Table 3.1. Tetrick et al. (2016) argued that incident response managers can use such classifications in several ways. First, the processes described in this table can be used to inform job and position descriptions, increasing their accuracy and facilitating employee selection. Second, this classification can inform performance evaluation and management tools, which in turn can foster performance feedback. Finally, the classification can be used to design team and MTS training programs that target particular sets of transition and action processes.

CSIRT and CSIR-MTS Outputs

Thus far, we have described the team level inputs and processes that can foster CSIRT and CSIR-MTS success. The inputs we specified at both levels set the foundation for the effective engagement of transition and action processes within and between component teams. In this section, we describe some markers of team and MTS effectiveness in CSIR.

Tetrick et al. (2016; see also Zaccaro et al., 2016) defined eight categories of cybersecurity performance metrics identified from interviews with incident response managers. The first refers to quantity indicators that include the number of incidents handled as well as the percentage of unsuccessful versus successful cyber-attacks and the number of viruses detected in a system. The second set refers to efficiency metrics such as speed and average time needed to triage threats and mitigate vulnerabilities. A third set of performance metrics contains indicators of a secure configuration and includes the "percentage of the system with no known severe vulnerabilities... [and] with security accreditations/certifications" (Tetrick et al., 2016, p. 74). This set also includes the number of devices that have acceptable antivirus software.

The fourth set of performance metrics derived by Tetrick et al. (2016) from manager interviews refers to incident handling capabilities, and includes related indicators such as the time needed to close detected system vulnerabilities and the amount of downtime caused by cyber-attacks and incidents. The amount of time needed to recover from such events and to remediate or reconfigure a system after an attack are included in a fifth set of metrics labeled recovery capability. Metrics associated with the overall reduction of system vulnerability, including the identification of weaknesses in the system constitute a sixth set of performance indicators. The final sets of metrics derived by Tetrick et al. include employee turnover statistics and changes in perceptions of an organization's reputation based on media accounts. Tetrick et al.

(2016) provide elaborated definitions and descriptions of these eight sets of metrics (see their Addendum 3.1, pp. 74–75).

The aforementioned metrics are fairly objective in nature. Researchers have also identified indicators of CSIRT and CSIR-MTS performance that are more subjective (Dalal et al., 2022; Tetrick et al., 2016; Zaccaro et al., 2016). These can include manager ratings of team and MTS performance as well as ratings against benchmarks from CSIR maturity frameworks (e.g., Stikvoort, 2015). Other performance outcomes include changes in team and MTS emergent states, such as viability, cohesion, trust, collective confidence, and shared knowledge (Marks et al., 2001; Tetrick et al., 2016). Taken together with the more objective metrics mentioned above, these outputs can be used to evaluate the effectiveness of processes in CSIRTs and CSIR-MTSs, as well determine how well these collectives are utilizing the inputs and resources available to them.

CSIRT AND CSIR-MTS CHALLENGES

In this final section of our chapter, we focus on two challenges typically confronting managers of incident response teams and MTSs. These include (1) attaining social maturity; and (2) managing MTS differentiation.

Attaining Social Maturity

Researchers in CSIR have offered dimensions to evaluate CSIRT maturity (e.g., Butkovic & Caralli, 2013; NCSC-NL, 2015; Stikvoort, 2015). Such maturity has been defined as:

> An indication of how well a team governs, documents, performs, and measures the CSIRT services. Is the CSIRT aware of the various processes and the required steps? Are these written down, shared, examined, and improved? Is it clear what the CSIRT authority and accountability is? Are there mechanisms to ensure that the CSIRT follows the formal processes and adequately serves its constituency? Are there mechanisms in place to constantly learn and improve?
>
> **(NCSC-NL, 2015, p. 2)**

While highly useful, these maturity models make minimal if any mention of the quality of collaboration and teamwork processes extant in CSIRTs and CSIR-MTSs (Tetrick et al., 2016).

Earlier, we described interactions and collective behavior that denote team and MTS processes used in mission accomplishment. How *well* cyber team and MTS members engage in these processes is called the *social maturity* of the cyber collective, which, according to Dalal and colleagues, consists of nine sets of collective skills and capabilities (Dalal et al., 2022; Tetrick et al., 2016): (1) communication; (2) information sharing; (3) collaborative problem solving; (4) shared cognition; (5) trust; (6) collective adaptation; (7) collective learning; (8) conflict management; and (9) effective collaboration triggering (see above sources for additional information about these elements of social maturity).

The challenge in fostering social maturity in incident response collectives is that managers typically focus most of their attention on developing the maturity of the cybersecurity technological and organizational infrastructure. While some CSIR maturity models have begun to include aspects of teamwork (e.g., Stikvoort, 2015), managers still tend to neglect the development of the nine aforementioned elements of social maturity. Such development begins with the hiring of incident response personnel and the staffing of CSIRT teams and MTSs. For example, managers rarely consider attributes such as preference for working in teams and collaboration skills in their hiring protocols for cyber analysts (Tetrick et al., 2016). Yet, such attributes are foundational individual skills for effective teamwork (Driskell et al., 2010). Likewise, managers need to establish performance evaluation systems that target not only task effectiveness, but also the nine elements of incident response social maturity. Finally, incident response training needs to focus not only on the "taskwork" (e.g., technical and analytical skills) of CSIR, but also the "teamwork" needed for effective team and MTS incident management (Salas et al., 2018). Tetrick et al. (2016) present several selection, performance evaluation, and training strategies for enhancing the social maturity of both CSIRTs and CSIR-MTSs.

Managing MTS Differentiation

A potentially serious problem in managing incident response MTSs may arise from high differentiation among the component teams in the MTS (Luciano et al., 2018). Luciano et al. (2018, p. 1067) defined MTS differentiation as "the degree of difference and separation between MTS component teams at a particular point in time." Earlier, we described external MTSs as those composed of component teams from different organizations. This aspect of the MTS structure can heighten the psychological distance existing among such teams. Moreover, such differentiation from external MTS structures can be exacerbated when component teams also come from different national cultures. According to Tetrick and colleagues (2016), managers and members of multicultural CSIR-MTSs describe significant challenges in communicating and collaborating with members of component teams from different cultures.

MTS differentiation can harm the development of social maturity in CSIRTs and CSIR-MTSs. Luciano et al. (2018) argued that greater differentiation in MTSs produces boundary enhancing forces whereby members of component teams emphasize the similarities within the team, the salience of team membership, and differences with other component teams in the system. The result can be stronger barriers to cross-team collaboration and communication. Indeed, in their examination of 40 failed MTSs, Campbell et al. (2022) found that members of component teams tend to focus more on within-team alignment behaviors (as opposed to between-team alignment), and this tendency was worse in external teams, especially in later stages of worsening crises.

Managers of highly differentiated CSIR-MTSs (because of functional, cultural, organizational, and geographic differences among component teams) can use several strategies to address the challenges such structures pose. First, at minimum, they need to increase their awareness of how their MTSs are structured and what differences exist among the component teams. The MTS mapping tools from Tetrick et al. (2016) that were described earlier can help managers identify their separate teams, perceive differences in interaction patterns, and define what variables may change these between-team interactions. Such tools can help managers maintain a focus on their MTS as a whole, as well as on different component teams.

Managers can also address challenges from differentiation by fostering a stronger MTS identity in addition to team identity (Zaccaro, et al., 2022). This includes emphasizing the distal goals of the MTS and how each and all component teams contribute to its accomplishment. Such efforts increase the salience of the MTS as a whole and may break down some of the boundary enhancing forces that can occur in highly differentiated MTSs. Managers can create linking mechanisms such as cross-functional teams or cross-boundary teams that focus on between-team alignment; such efforts can foster create shared cognition across component teams in an MTS, especially when paired with cross-training, where members of different teams become more aware of the functions of other teams (Luciano et al., 2018). Tetrick et al. (2016) also proposed that managers use communication charters that specify the protocols and expectations for cross-boundary information sharing and collaboration. These strategies all have the goal of reducing boundary-enhancing forces among component teams and increasing boundary spanning activities.

Summary

CSIRTs often fail because of insufficient attention given to developing collective skills in teamwork and collaboration. Moreover, even less attention is given to when teams are operating within closely connected networks of teams to reach collective goals. In this chapter, we describe the elements of CSIRTs and MTSs, including (a) inputs that contribute to effective processes at each level, (b) the team and MTS processes specific to CSIR, and (c) the range of potential outputs of these processes. We also elucidated two key challenges to incident response team and MTS effectiveness: attaining social maturity and managing MTS differentiation. We hope this chapter fosters a greater focus in research and practice on the social maturity of such teams and MTSs. We also hope the concepts we offer in this chapter will point CSIR managers toward strategies and tools for improving their CSIRTs and CSIR-MTSs.

REFERENCES

Ahmad, A., Maynard, S. B., Desouza, K. C., Kotsias, J., Whitty, M. T., & Baskerville, R. L. (2021). How can organizations develop situation awareness for incident response: A case study of management practice. *Computers & Security, 11*, 102122.

Alberts, C., Dorofee, A., Killcrece, G., Ruefle, R., & Zajicek, M. (2004, October). *Defining incident management processes for CSIRTs: A work in progress* (CMU/SEI- 2004-TR-015). Retrieved from https://apps.dtic.mil/sti/pdfs/ADA453378.pdf..

Alliger, G. M., Cerasoli, C. P., Tannenbaum, S. I., & Vessey, W. B. (2015). Team resilience: How teams flourish under pressure. *Organizational Dynamics, 44*(3), 176–184. https://doi.org/10.1016/j.orgdyn.2015.05.003

Andrews, R. W., Lilly, J. M., Srivastava, D., & Feigh, K. M. (2023). The role of shared mental models in human-AI teams: A theoretical review. *Theoretical Issues in Ergonomics Science, 24*(2), 129–175. https://doi.org/10.1080/1463922X.2022.2061080.

Bernerth, J. B., Beus, J. M., Helmuth, C. A., & Boyd, T. L. (2023). The more the merrier or too many cooks spoil the pot? A meta-analytic examination of team size and team effectiveness. *Journal of Organizational Behavior, 44*(8), 1230–1262. https://doi.org/10.1002/job.2708

Bhaskar, R. (2005). A proposed integrated framework for coordinating computer security incident response teams. *Journal of Information Privacy and Security, 1*(3), 3–17.

Breuer, C., Hüffmeier, J., & Hertel, G. (2016). Does trust matter more in virtual teams? A meta-analysis of trust and team effectiveness considering virtuality and documentation as moderators. *Journal of Applied Psychology, 101*(8), 1151–1177. https://doi.org/10.1037/apl0000113

Breuer, C., Hüffmeier, J., Hibben, F., & Hertel, G. (2020). Trust in teams: A taxonomy of perceived trustworthiness factors and risk-taking behaviors in face-to-face and virtual teams. *Human Relations, 73*(1), 3–34. https://doi.org/10.1177/0018726718818721

Butkovic, M. J., & Caralli, R. A. (2013). *Advancing cybersecurity capability measurement using the CERT (registered trademark) - RMM maturity indicator lead scale* (No. CMU/SEI-2013-TN-028). Software Engineering Institute.

Campbell, D. J. (1988). Task complexity: A review and analysis. *The Academy of Management Review, 13*(1), 40–52. https://doi.org/10.2307/258353

Campbell, D. T. (1958). Common fate, similarity, and other indices of the status of aggregates of persons as social entities. *Behavioral Science, 3*, 14–25.

Campbell, L. N. P. (2024). *Theoretical foundation and qualitative analysis of resilience in multiteam systems* [Unpublished doctoral dissertation]. George Mason University.

Campbell, L. N. P., Torres, E. M., Zaccaro, S. J., Zhou, S., Hedrick, K. N., Wallace, D. M., Luning, C. R., & Zakzewski, J. E. (2022). Examining multiteam systems across context and type: A historiometric analysis of failed MTS performance. *Frontiers in Psychology: Organizational Psychology, 13*(813624), 1–22.

Checklist Incident Priority. (n.d.). IT Process Maps. http://wiki.en.it-processmaps.com/index.php/Checklist_Incident_Priority.

Chen, T. R., Shore, D. B., Zaccaro, S. J., Dalal, R. S., Tetrick, L. E., & Gorab, A. K. (2014). An organizational psychology perspective to examining computer security incident response teams. *IEEE Security & Privacy, 12*, 61–67.

Chickowski, E. (2019, September 2). Every hour SOCs run, 15 minutes are wasted on false positives. https://securityboulevard.com/2019/09/every-hour-socs-run-15-minutes-are-wasted-on-false-positives/

Cichonski, P., Millar, T., Grance, T., & Scarfone, K. (2012). *Special Publication 800-61 Revision 2; Computer Security Incident Handling Guide.* NIST, US Department of Commerce. http://dx.doi.org/10.6028/NIST.SP.800-61r2

Cuevas, A., Hogan, E., Hanan Hibshi, H., & Christin, N. (2021). Observations from an online security competition and its implications on crowdsourced security. arXiv:2204.12601 **[cs.CR]**. https://arxiv.org/abs/2204.12601 (January 22, 2025).

Dalal, R. S., Bolunmez, B., Tomasseti, A. J., & Sheng, Z. (2016). Escalation: An understudied team decision-making structure. In S. J. Zaccaro, R. S. Dalal, L. E. Tetrick, & J. A. Steinke (Eds.), *Psychosocial dynamics of cyber security* (pp. 74–91). Routledge.

Dalal, R. S., Howard, D. J., Bennett, R., Posey, C., Zaccaro, S. J., & Brummel, B. J. (2022). Organizational science and cybersecurity: Abundant opportunities for research at the interface. *Journal of Business and Psychology, 37*(1), 1–29.

Davison, R. B., Hollenbeck, J. R., Barnes, C. M., Sleesman, D. J., & Ilgen, D. R. (2012). Coordinated action in multiteam systems. *Journal of Applied Psychology, 97*(4), 808–824.

DeChurch, L. A., Burke, C. S., Shuffler, M. L., Lyons, R., Doty, D., & Salas, E. (2011). A historiometric analysis of leadership in mission critical multiteam environments. *The Leadership Quarterly*, *22*(1), 152–169.

DeChurch, L. A., & Marks, M. A. (2006). Leadership in multiteam systems. *Journal of Applied Psychology*, *91*(2), 311–329. https://doi.org/10.1037/0021-9010.91.2.311

DeChurch, L. A., & Zaccaro, S. J. (2013). Innovation in scientific multiteam systems: Confluent and countervailing forces. *Prepared for the national research council's workshop on science team dynamics and effectiveness*. https://sites.nationalacademies.org/cs/groups/dbassesite/documents/webpage/dbasse_083773.pdf

Dennis, A. R., Lakhiwal, A., & Sachdeva, A. (2023) AI agents as team members: Effects on satisfaction, conflict, trustworthiness, and willingness to work with. *Journal of Management Information Systems*, *40*(2), 307–337. https://doi.org/10.1080/07421222.2023.2196773

Dorofee, A., Killcrece, G., Ruefle, R., & Zajicek, M. (2007). Incident management capability metrics: Version 0.1. (CMU/SEI-2007-TR-008). https://apps.dtic.mil/sti/tr/pdf/ADA468688.pdf

Driskell, J. E., Salas, E., & Hughes, S. (2010). Collective orientation and team performance: Development of an individual differences measure. *Human Factors*, *52*, 316–328.

Equifax (2017, September 7). Equifax announces cybersecurity incident involving consumer information, [Press release]. https://investor.equifax.com/news-events/press-releases/detail/240/equifax-announces-cybersecurity-incident-involving-consumer. *For example, Equifax suffered a cybersecurity breach on its Automated Consumer Interview System on May 13, 2017, but it was not detected until July 29, 2017 (Equifax, 2017).*

Fleishman, E. A., & Zaccaro, S. J. (1992). Taxonomic classifications of team tasks. In R. W. Swezey & E. Salas (Eds.), *Teams: Their training and performance* (pp. 31–56). ABLEX.

Gilson, L. L., Maynard, M. T., Jones Young, N. C., Vartiainen, M., & Hakonen, M. (2015). Virtual teams research: 10 years, 10 themes, and 10 opportunities. *Journal of Management*, *41*(5), 1313–1337. https://doi.org/10.1177/0149206314559946

Grobler, M., & Bryk, H. (2010). *2010 Information Security for South Africa*, Johannesburg, South Africa, pp. 1–6. https://doi.org/10.1109/ISSA.2010.5588307.

Hackman, J. R., & Morris, C. G. (1975). Group tasks, group interaction process, and group performance effectiveness: A review and proposed integration. In L. Berkowitz (Ed.),. *Advances in experimental social psychology* (Vol. 8, pp. 45–99). Academic Press.

Harris-Watson, A. M., Larson, L. E., Lauharatanahirun, N., DeChurch, L. A., & Contractor, N. S. (2023). Social perception in Human-AI teams: Warmth and competence predict receptivity to AI teammates. *Computers in Human Behavior*, *145*, 107765.

Hollenbeck, J. R., Beersma, B., & Schouten, M. E. (2012). Beyond team types and taxonomies: A dimensional scaling conceptualization for team description. *The Academy of Management Review*, *37*(1), 82–106.

Ilgen, D. R., Hollenbeck, J. R., Johnson, M., & Jundt, D. (2005). Teams in organizations: From input-process-output models to IMOI models. *Annual Review of Psychology*, *56*(1), 517–543.

Jose, I., LaPort, K., & Trippe, D. M. (2016). Requisite attributes for cybersecurity personnel and teams: Cyber risk mitigation through talent management. In S. J. Zaccaro, R. D. Dalal, L. E. Tetrick & J. A. Steinke (Eds.), *The psychosocial dynamics of cyber security* (pp. 167–193). Routledge.

Kabanov, I., & Madnick, S. E. (2020). A systematic study of the control failures in the Equifax cybersecurity incident. MIT Sloan Research Paper No. 2020-19. https://ssrn.com/abstract=3957272

Kam, H.J., Ormond, D. K., Menard, P., & Crossler, R. E. (2022). That's interesting: An examination of interest theory and self-determination in organisational cybersecurity training. *Information Systems Journal*, *32*(4), 888–926. https://doi.org/10.1111/isj.12374

Killcrece, G., Kossakowski, K.P., Ruefle, R., & Zajicek, M. (2003). State of the practice of computer security incident response teams (CSIRTs). Carnegie Mellon University Software Engineering Institute Technical Report. https://resources.sei.cmu.edu/asset_files/TechnicalReport/2003_005_001_14204.pdf

Klimoski, R., & Murray, J. (2016). Cyber security executive leadership. In S. J. Zaccaro, R. D. Dalal, L. E. Tetrick, & J. A. Steinke (Eds.), *The psychosocial dynamics of cyber security* (pp. 135–166). Routledge.

Kozlowski, S. W. J., & Ilgen, D. R. (2006). Enhancing the effectiveness of work groups and teams. *Psychological Science*, *7*, 77–124.

Lane, J. N., Leonardi, P. M., Contractor, N. S., & DeChurch, L. A. (2024). Teams in the digital workplace: Technology's role for communication, collaboration, and performance. *Small Group Research*, *55*(1), 139–183. https://doi.org/10.1177/10464964231200015

Litman, J. A. (2008). Interest and deprivation factors of epistemic curiosity. *Personality and Individual Differences*, *44*, 1585–1595.

Lorson, F., Fügener, A., & Hübner, A. (2023). New teammates in the warehouse: Human interactions with automated and robotized systems. *IISE Transactions*, *55*(5), 536–553. https://doi.org/10.1080/24725854.2022.2072545.

Luciano, M. M., DeChurch, L. A., & Mathieu, J. E. (2018). Multiteam systems: A structural framework and meso-theory of system functioning. *Journal of Management*, *44*(3), 1065–1096.

Marks, M. A., Mathieu, J. E., & Zaccaro, S. J. (2001). A temporally based framework and taxonomy of team processes. *Academy of Management Review*, *26*(3), 356–376.

Mathieu, J. E., Gallagher, P. T., Domingo, M. A., & Klock, E. A. (2019). Embracing complexity: Reviewing the past decade of team effectiveness research. *Annual Review of Organizational Psychology and Organizational Behavior*, *6*, 17–46. https://doi.org/10.1146/annurev-orgpsych-012218-015106

Mathieu, J. E., Hollenbeck, J. R., van Knippenberg, D., & Ilgen, D. R. (2017). A century of work teams in the *Journal of Applied Psychology*. *Journal of Applied Psychology*, *102*(3), 452–467. https://doi.org/10.1037/apl0000128

Mathieu, J. E., Marks, M. A., & Zaccaro, S. J. (2001). Multiteam systems. In N. Anderson, D. S. Ones, H. K. Sinangil, & C. Viswesvaran (Eds.), *Organizational psychology: Vol. 2. Handbook of industrial, work and organizational psychology* (2nd ed., pp. 289–313). Sage.

Maxam, W. P. III, & Davis, J. C. (2024). An interview study on third-party cyber threat hunting processes in the US Department of Homeland Security. https://www.usenix.org/system/files/usenixsecurity24-maxam.pdf

McGrath, J. E. (1964). *Social psychology: A brief introduction*. Holt, Rinehart and Winston.

Milani, A. M. P., Starr, A., Hill, S., Curtis, C., Anderson, N., Moreno-Lumbreras, D., & Storey, M. (2024). Fuzzy to clear: Elucidating the threat hunter cognitive process and cognitive support needs. arXiv:2408.04348 [**cs.CR**]. https://doi.org/10.48550/arXiv.2408.04348

Morgeson, F. P. (2005). The external leadership of self-managing teams: Intervening in the context of novel and disruptive events. *Journal of Applied Psychology*, *90*(3), 497–508. https://doi.org/10.1037/0021-9010.90.3.497

Mumford, M. D., Hester, K. S., & Robledo, I. C. (2012). Creativity in organizations: Importance and approaches. In M. D. Mumford (Ed.), *Handbook of organizational creativity* (pp. 3–16). Academic Press. https://doi.org/10.1016/B978-0-12-374714-3.00001-X

Mumford, M. D., Mobley, M. I., Reiter-Palmon, R., Uhlman, C. E., & Doares, L. M. (1991). Process analytic models of creative capacities. *Creativity Research Journal*, *4*(2), 91–122. https://doi.org/10.1080/10400419109534380

Murase, T., Carter, D. R., DeChurch, L. A., & Marks, M. A. (2014). Mind the gap: The role of leadership in multiteam system collective cognition. *The Leadership Quarterly, 25*(5), 972–986. https://doi.org/10.1016/j.leaqua.2014.06.003

National Cyber Incident Scoring System. (n.d.) https://www.cisa.gov/sites/default/files/2023-01/cisa_national_cyber_incident_scoring_system_s508c.pdf

NCSC, Netherlands (NCSC-NL). (2015). CSIRT Maturity Kit: *A step-by-step guide towards enhancing CSIRT Maturity.* https://www.ncsc.nl/binaries/ncsc/documenten/publicaties/2019/mei/01/csirt-maturity-kit/CSIRT_MK_guide.pdf

Obama, B. H. (2009). Remarks by the President on securing our nation's cyber infrastructure. Retrieved January 20, 20025, from https://obamawhitehouse.archives.gov/the-press-office/remarks-president-securing-our-nations-cyber-infrastructure.

Räetze, S., Duchek, S., Maynard, M. T., & Wohlgemuth, M. (2022). Resilience in organization-related research: An integrative conceptual review across disciplines and levels of analysis. *Journal of Applied Psychology, 107*(6), 867–897. https://doi.org/10.1037/apl0000952

Reinhardt, W., Schmidt, B., Sloep, P., & Drachsler, H. (2011). Knowledge worker roles and actions—Results of two empirical studies. *Knowledge and Process Management, 18*(3), 150–174.

Richberg, J. (2022, October 17). *Three Ways to Improve Teamwork in Cybersecurity.* Forbes. https://www.forbes.com/councils/forbestechcouncil/2022/10/17/three-ways-to-improve-teamwork-in-cybersecurity/

Salas, E., Reyes, D. L., & McDaniel, S. H. (2018). The science of teamwork: Progress, reflections, and the road ahead. *American Psychologist, 73*(4), 593–600. https://doi.org/10.1037/amp0000334

Seeber, I., Bittner, E. A. C., Briggs, R. O., de Vreede, T., de Vreede, G., Elkins, A., & Söllner, M. (2020). Machines as teammates: A research agenda on AI in team collaboration. *Information & Management, 57,* 103174. https://doi.org/10.1016/j.im.2019.103174

Shaikh, F. A., & Siponen, M. (2023). Information security risk assessments following cybersecurity breaches: The mediating role of top management attention to cybersecurity. *Computers & Security, 124,* 102974. https://doi.org/10.1016/j.cose.2022.102974

Software Engineering Institute (SEI). (2014). Software assurance for executives: Definitions. https://resources.sei.cmu.edu/asset_files/EducationalMaterial/2014_011_001_81821.pdf

Sparks, J. (2015, June). *Effective team leadership and process improvement for network security operators.* Presentation at the 27th Annual FIRST Conference, Berlin, Germany. https://www.first.org/resources/papers/conf2015/first_2015_-_sparks-jeremy-_effective_team_leadership_20150630.pdf

Stikvoort, D. (2015). *SIM3: Security incident management maturity model.* https://opencsirt.org/wp-content/uploads/2019/12/SIM3-mkXVIIIc.pdf

Stoverink, A. C., Kirkman, B. L., Mistry, S., & Rosen, B. (2020). Bouncing back together: Toward a theoretical model of work team resilience. *Academy of Management Review, 45*(2), 395–422. https://doi.org/10.5465/amr.2017.0005

Tesluk, P., Mathieu, J. E., Zaccaro, S. J., & Marks, M. (1997). Task and aggregation issues in the analysis and assessment of team performance. In M. T. Brannick, E. Salas, & C. Prince (Eds.), *Team performance assessment and measurement: Theory, methods, and applications* (pp. 197–224). LEA.

Tetrick, L. E., Zaccaro, S. J., Dalal, R. S., Steinke, J. A., Repchick, K. M., Winslow, C. J., … Schrader, S. W. (2016). *Improving social maturity of cybersecurity incident response teams.* George Mason University. http://calctraining2015.weebly.com/the-handbook.html

Torres, E. M., Wallace, D. M., Zaccaro, S. J., & Dubrow, S. (2021). Deconstructing multiteam system action: Development and content validation of a multilevel multiteam system action taxonomy. *Human Performance, 34*(3), 189–216. https://doi.org/10.1080/08959285.2021.1922909

Wood, R. E. (1986). Task complexity: Definition of the construct. *Organizational Behavior and Human Decision Processes, 37*(1), 60–82. https://doi.org/10.1016/0749-5978(86)90044-0

Zaccaro, S. J., Hargrove, A., Chen, T. R., Repchick, K., & McCausland, T. (2016). A comprehensive multilevel taxonomy of cybersecurity incident response performance. In S. J. Zaccaro, R. D. Dalal, L. E. Tetrick, & J. A. Steinke (Eds.), *The psychosocial dynamics of cyber security*. Routledge.

Zaccaro, S. J., Marks, M. A., & DeChurch, L. A. (2012). Multiteam systems: An introduction. In S. J. Zaccaro, M. A. Marks, & L. DeChurch (Eds.), *Multiteam systems: An organization form for dynamic and complex environments* (pp. 3–31). Routledge.

Zaccaro, S. J., Weis, E. J., Zakzewski, J. E., & Hilton, R. (2022). Building resilient teams. In P. J. Sweeney, M. D. Mathews, P. B. Lester, S. T. Hannah, & B. J. Reed (Eds.), *Leadership in dangerous situations: A handbook for the armed forces, emergency services, and first responders*. Naval Institute Press.

4 Enhancing Cybersecurity through Teamwork

Developing Effective Security Operation Center Teams

Lila Berger, Maha Khalid, Rylee Linhardt, Gabriela Fernández Castillo, and Eduardo Salas

INTRODUCTION

As technology continues to advance, awareness and knowledge regarding the potential risks and threats to cybersecurity must simultaneously keep pace. Equally critical is the preparation of cybersecurity teams with the requisite skills to effectively mitigate and defend against evolving threats. These teams, situated within security operations centers (SOCs), may operate either internally or externally to an organization (Agyepong et al., 2020). An SOC is a team consisting primarily of security professionals whose core roles and responsibilities encompass the identification, analysis, response, reporting, and even prevention of cybersecurity issues (Zimmerman, 2014; Kokulu et al., 2019). These individuals serve as the "first responders" during cybersecurity breaches or attacks, necessitating collaborative teamwork to resolve such incidents, which can often extend over several days or weeks. In any given year, an SOC must monitor and resolve thousands of threats and potential vulnerabilities (Zimmerman, 2014). Each vulnerability, threat, or scenario necessitates an appropriate and rapid response from an SOC team. Due to the fast-paced and critical nature of this work, SOC teams operate in environments constituted by extreme levels of stress, in addition to volatility, uncertainty, complexity, and ambiguity (VUCA), where wrong decisions can have drastic consequences; the literature labels such an environment an extreme or VUCA environment (Driskell et al., 2018). It is therefore especially necessary for these teams to establish and utilize a robust team framework to promote overall team effectiveness, among other critical teamwork outcomes. Despite this necessity, SOC teams often lack an adequate understanding regarding the teamwork competencies and outcomes necessary for work in such environments, which can be problematic for SOC success (Simonson et al., 2020, as cited in Sinlapanuntakul et al., 2022).

It is critical for SOC teams operating under such extreme conditions to rely on fundamental teamwork principles to aid in team development. Foundational team

DOI: 10.1201/9781003413707-5

science processes and emergent states shown to aid in the development of an effective team in extreme environments include the development of a shared mental model (SMM), psychological safety, shared vision, cooperation and coordination, situation awareness, and mutual trust, which have been shown to be important for cybersecurity teams (Steinke et al., 2015; Sinlapanuntakul et al., 2022), as well as generally for developing effective teams working in extreme environments (Dietz et al., 2017; Paoletti et al., 2020). Additionally, cognitive bias has been shown to impact both decision-making in cybersecurity contexts (Lemay & Leblanc, 2018) and also for the development of effective teamwork in extreme environments (Antonacci et al., 2021). Similarly, when working in such environments, workload demands have also been suggested to impact cybersecurity-related performance (reduction in threat detection, etc.; Paul & Dykstra, 2017) as well as general team effectiveness in extreme environments (Bowers et al., 1997; Hagemann et al., 2023). Hence, these processes and emergent states have grounding in both the cybersecurity and extreme environments literature, and therefore serve as essential mediators for the development of team outcomes such as team effectiveness, learning and adaptability, resilience, and well-being in the cybersecurity context.

Another critical mediating mechanism to consider when developing an effective SOC team is decision-making. The operational environment of SOC teams demands quick decision-making under heightened stress and with the utilization of limited data. Within security operation centers, gaining an understanding of decision-making processes and models can offer insights into developing more effective strategies, anticipating potential threats, and fostering proactive and resilient security protocols. The literature on *Naturalistic Decision Making* (NDM) can provide a lens through which we can examine the evolving landscape of cybersecurity, as there is limited research in this area. Research on NDM began in the 1980s in an attempt to understand how individuals make decisions in the real-world (Klein, 2008). Rather than taking a domain-independent approach, NDM focuses on a knowledge-based approach that considers decision-makers who have substantial experience. In this context, this would include the individuals working in security operation centers who make high-stakes decisions under time pressure and VUCA conditions consistently. Models situated within NDM such as Recognition-Primed Decision-Making (RPD; see *Framework Mediating Mechanisms* for more on RPD) can help to mediate the development of effective teams in the cybersecurity context, among other important team outcomes.

Leveraging insights and theory from interdisciplinary literature can offer valuable guidance for informing SOC team development. This chapter aims to utilize the extreme environments, decision-making, cybersecurity, and team science literature to develop a framework describing cybersecurity teams' critical mediating mechanisms and their corresponding team outcomes. The field of cybersecurity could benefit from the incorporation and integration of knowledge translation from this literature in order to enhance team effectiveness under stressful conditions, among other outcomes. Team constructs necessary for the development of effective teams (i.e., SMMs, psychological safety, shared vision, cooperation and coordination, SA, mutual trust) and other cognitive factors (i.e., cognitive bias and workload) will be sourced from this literature and translated to the cybersecurity domain.

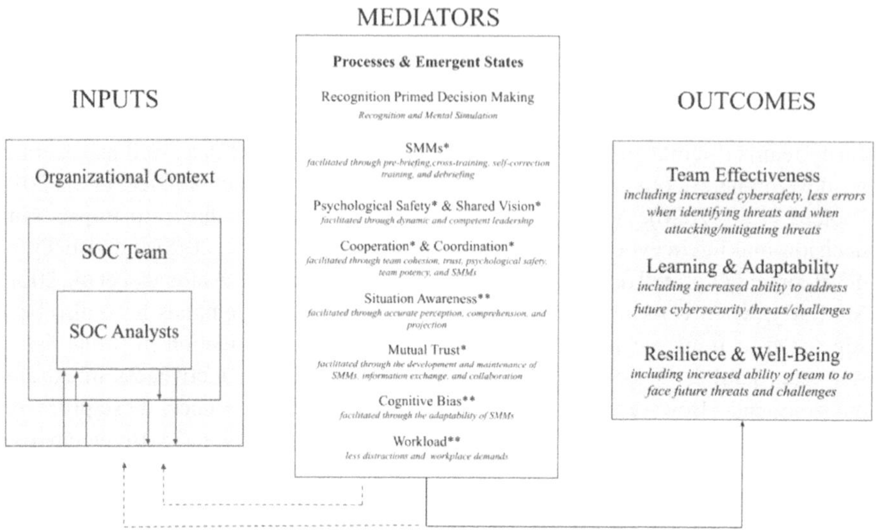

FIGURE 4.1 Security Operation Center (SOC) framework.

Note: Input-mediator-output-input framework, adapted from Grossman et al. (2017), emphasizing the components included in this chapter.

* indicates team-level variables, while ** indicates this variable can exist at both the individual-level and at the team-level. Absence of * indicates this variable exists only at the individual-level.

FRAMEWORK

To foster a better understanding of the intricacies of how SOC teams may operate and utilize critical processes and emergent states to facilitate better outcomes, teams research has historically turned to what is known as the input-mediator-output (IPO) model (Hackman, 1987; McGrath, 1964, as cited in Grossman et al., 2017), which provides a theoretical framework to help foster a deeper understanding of the complexities that may be involved in the facilitation of various team outcomes. We have utilized this literature to translate these teamwork principles necessary for the development of an effective team to create a framework (Figure 4.1), which serves to provide a structure for the insights detailed in this chapter; aiding in the attainment of improved outcomes for teams working in the cybersecurity domain.

FRAMEWORK INPUTS

Inputs are the first step in this framework, and these components may include certain team member attributes or environmental conditions (Grossman et al., 2017). In this framework (Figure 4.1), we provide domain-specific contextual inputs surrounding this occupational environment, which involves teams of SOC analysts working to mitigate and reduce cybersecurity threats and attacks in various organizational contexts.

FRAMEWORK MEDIATING MECHANISMS

The meditating mechanisms of this framework involve the processes and emergent states critical to the fostering of the previously mentioned outcomes. Mediators facilitate the previously mentioned outcomes to various degrees, and all have evidence-based backings in the psychological sciences literature. The mediating mechanisms and their facilitators are detailed in the following sections.

Recognition Primed Decision-Making

Arguably one of the most important elements of an SOC team's job includes the process of decision-making, a key component to our framework. Decision-makers have expanded the decision-making process to include a preliminary stage involving the perception and recognition of situations, along with the generation of appropriate responses (Klein, 2008). Rather than selecting from a predetermined list of options, this approach capitalizes on using the decision-maker's previously held scripts, schema, and mental models. Utilizing such mechanisms differentiates an expert from a novice. This model is called the RPD model of NDM (Ross et al., 2004).

In studies examining human cognition, cognitive processes related to decision-making have been classified into two systems: System 1 (fast and unconscious) and System 2 (slow and deliberate; Salas et al., 2010). A solely intuitive strategy when making high-risk decisions in the cybersecurity context would be too risky, whereas a deliberative and analytical approach would be too slow. As such, RPD bridges the gap by blending intuition and analysis, providing decision-makers with a balanced and effective approach to tackling cybersecurity challenges.

According to the RPD model, there are two processes that relate to expertise acquisition – *recognition* and *mental simulation* (see Figure 4.2; Klein, 2008). Recognition includes cue recognition, generation of expectancies, identification of relevant goals, and recognition of typical actions (Ross et al., 2004). A *cue* refers to a direct aspect of a situation that allows the individual to think that a larger pattern likely exists in a situation (Ross et al., 2004). In that given pattern, *expectancies* allow the individual to develop a story of the current state and how other related scenarios could unfold. Then, the identification of relevant goals allows the decision-maker to align their actions with the overall objectives of the task, and then recognition of typical actions follow. These procedures in the recognition phase allow experiences in a given situation to get integrated into one's mental model (Ross et al., 2004).

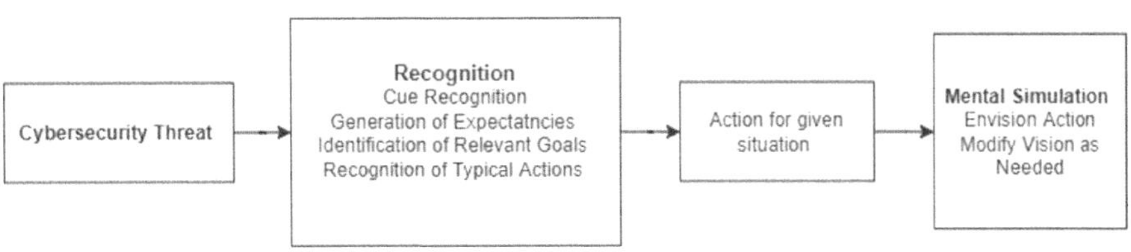

FIGURE 4.2 Processes of the RPD model given cybersecurity threat. (Adapted from file:///Users/bergerl1/Downloads/78Ross et al. 2004.)

Understanding patterns and applying them involve two distinct forms of knowledge. After an examination of the situation, an expert is able to identify a course of action that has proven effective in previously similar situations. This is where mental simulations come into play, constituting the second process in the RPD model. Mental simulations allow decision-makers to envision the unfolding of the chosen action in their mind, simulating how the decision interacts with other features of the current situation over time. In other words, skilled decision-makers are not required to compute choices, but rather possess the ability to instinctively respond to a situation based on their perceptual understanding of it (Klein & Crandall, 2018).

In situations such as the one where a cybersecurity threat is present, effective teams may be more likely to construct complete and accurate mental models, have superior schemata, have richer action repertories, and use appropriate decision rules (Lipshitz & Shaul, 1997). The knowledge-based performance required in RPD is characterized by the generation of new rules in a novel situation. Professionals in the cybersecurity domain acquire a robust conceptual grasp of the intricacies within their domain. However, knowledge alone does not guarantee expert-level proficiency. Achieving expertise requires repetitive engagement in tasks that can lead to automation of certain cognitive processes that allows for swift and accurate execution with reduced mental workload (Ross et al., 2004). As the fundamental elements of similar scenarios become automatic, experts and effective teams can navigate and manipulate more intricate models without an increase in cognitive effort. This heightened level of adaptability allows experts to effectively respond to evolving threats and challenges.

As SOC teams are called upon to address complex decision-making tasks, *macrocognition in teams* is of increasing importance. Derived from the NDM literature, macrocognition in teams describes a broad range of cognitive processes that allows for the generation or adaptation of rules in novel situations (Figure 4.1; Fiore et al., 2010). Specifically, the process of transforming internalized knowledge into externalized team knowledge is done through individual and team knowledge-building processes (Fiore et al., 2010). Developing macrocognition in teams requires (1) individual knowledge building, (2) team knowledge building, (3) externalized team knowledge, and (4) team problem-solving (Fiore et al., 2010).

The construction of individual knowledge requires engaging in activities carried out at the individual level to enhance one's own understanding in a given context. For instance, this knowledge acquisition can involve activities such as reading or engaging with a system (e.g., utilizing the RPD model in Figure 4.2). Once this knowledge is synthesized, team knowledge building can occur, which requires specific measures be taken at the interindividual level to foster knowledge development collectively. Subsequently, connections can be established through externalized team knowledge, in which team members work together using information and agreed-upon ideas to solve problems. Finally, the outcome is team problem-solving, which evaluates the quality of the team's solution or plans by checking if they meet all the criteria for solving the problems (Fiore et al., 2010).

Delving into the decision-making processes within SOCs can provide insights for crafting more effective strategies, anticipating threats, and enhancing security protocols. The NDM paradigm has previously focused on various domains including aviation, aerospace, energy production, defense, nuclear, medicine, oil and gas, and rail (Gore et al., 2018). Further research is needed to deepen our understanding of decision-making mechanisms given the complexity of domain-specific cybersecurity challenges, often characterized by high-stakes decision-making under time pressure and uncertain conditions. By enhancing our understanding, these teams can navigate complex scenarios with heightened effectiveness, adaptability, learning and resilience, which can foster a greater sense of well-being among members.

Shared Mental Models (SMMs)

SMMs are vital to increasing team effectiveness (Figure 4.1), as well as for cybersecurity teams more generally (Steinke et al., 2015), and can be defined as the emergent state involving the shared understanding among team members and team environmental elements (Marks et al., 2001; Mohammed et al., 2010). As such, SMMs have a significant impact on team performance as well as other teamwork processes such as coordination, adaptability, and creativity, to name a few (Müller and Antoni, 2022; Santos & Passos, 2013; Santos et al., 2015). SOC teams should strive to promote and develop SMMs among their teams, which can be fostered through pre-briefings, cross-training, guided team self-correction training, and debriefings (Steinke et al., 2015).

Pre-briefings can serve to improve team effectiveness by reducing communication inefficiencies and contribute to a team's SMM development (Steinke et al., 2015). Cross-training is another empirically-validated method for improving a team's SMM (Steinke et al., 2015). Specifically, cross-training allows for the team to have a more *accurate* mental model, hence fostering better comprehension regarding the roles, responsibilities, and timing for each team member (Steinke et al., 2015). There are various methods for cross-training, including watching, listening, and performing other members' roles within the team, all which have been shown to be effective and beneficial to not only the development of SMMs with a team, but also general team performance and effectiveness (Steinke et al., 2015). Self-correction training involves offering direction to members regarding the specific areas that can be improved or corrected related to team performance; this method has also been demonstrated to be effective for teams in extreme environments (Smith-Jentsch et al., 2008). Finally, similar to pre-briefings, debriefings are another popular and widely-used method for promoting a team's development of a SMM, among other benefits (Smith-Jentsch et al., 2008; Tannenbaum & Greilich, 2023). Debriefing refers to the practice of introspection regarding elements such as actions, challenges, achievements, areas for improvement, as well as upcoming plans and goals (Tannenbaum & Greilich, 2023). Consistent debriefing within a team has been shown to promote better team outcomes (Tannenbaum & Greilich, 2023), and therefore, SOC teams should also utilize debriefs to aid in the development of SMMs, among other benefits (Smith-Jentsch et al., 2008).

Psychological Safety and Shared Vision

Psychological safety is another important team emergent state for fostering desired team outcomes (Figure 4.1). Psychological safety can be defined as the belief among team members that it is safe to take interpersonal risks (Edmondson, 1999). This group-level phenomenon (Edmondson & Lei, 2014) is reliant upon both the development and maintenance of a shared vision as well as leadership (Figure 4.1). A shared vision is defined in the literature as an individual's comprehension of an organization's objectives, purpose, and goals (Eldor, 2020). Shared vision within a team can allow for the promotion of various positive outcomes, such as motivating the team and creating a shared purpose for their work (Bisbey et al., 2021). In the same vein, the *absence* of a shared vision within a team can inadvertently lead to undesirable outcomes such as poor communication and a lack of team cohesion (Bammer, 2018) that may impact general team performance and effectiveness (Klasmeier & Rowold, 2020).

Shown to impact the development of a shared vision within a team and psychological safety (Ou et al., 2018; Bisbey et al., 2021), leadership is a distinct competency (Bisbey et al., 2021), and competent leaders should set the team's course or direction, oversee team functioning, and enhance the team's capacity for self-management (Zaccaro et al., 2009). Teams possessing leaders with qualities such as dynamism and altruism are high-performing and more effective altogether (Wing, 2005).

To promote psychological safety within team structures, it is important that team members both understand and consider the team's mission above their own personal objectives (Driskell & Salas, 1992). Throughout the team science literature, psychological safety has been demonstrated to significantly impact many teamwork processes and outcomes. For example, teams with increased psychological safety have been shown to more efficiently learn and engage in information sharing (Cauwelier et al., 2019; Kim et al., 2020). SOC teams should encourage a workplace team culture in which all team members feel safe to voice their concerns, regardless of their occupational role or status. Psychological safety has been directly linked to performance and team effectiveness outcomes in the workplace (Edmondson & Lei, 2014; Newman et al., 2017).

Cooperation and Coordination

The next mediating processes in this framework are cooperation and coordination (Figure 4.1). Cooperation can be defined as the individual perspectives and beliefs that team members contribute to the group, while coordination refers to the requisite behaviors necessary for team effectiveness (Tannenbaum & Salas, 2021). Both cooperation and coordination have been shown throughout the literature to promote better team outcomes such as adaptability, resilience, and learning (Traylor et al., 2023). To promote cooperation and coordination among a team, team cohesion and trust are vital components, as well as psychological safety and team potency (Tannenbaum & Salas, 2021; Traylor et al., 2023). *Team trust and cohesiveness* become critical elements when teams are faced with unforeseen events and have to make decisions quickly and efficiently, both vital components of daily operations for SOC teams.

Psychological safety is also essential for cooperation and coordination, as members feel secure when voicing their opinions and concerns, hence allowing the team to coordinate their actions and cooperate with one another to make a consensual decision. Finally, *team potency*, marked by the overarching perception of a team's capacity to execute tasks across diverse situations (Bell et al., 2018), is another critical component of cooperation and coordination. SOC teams should feel prepared to balance confidence in their team with caution, and have the ability to accurately assess their knowledge, skills, and attributes (KSAs) when faced with daily challenges in the workplace. They must not only coordinate effectively with their fellow team members, but also be receptive to modifications or improvements in the team's KSAs as they encounter new challenges or need to adapt to the current situation. In teams marked by optimal coordination and collaboration, even if every member is not necessarily an "expert," teams can strategically utilize both coordination and cooperation to synchronize their efforts, ultimately attaining performance akin to that of experts. Hence, these two fundamental principles are necessary to help improve team outcomes, regardless of domain.

Situation Awareness

Teams of cybersecurity experts are employed to design and implement effective defense methodologies that neutralize a cyberthreat meant to damage an organization or entity. When teams are resolving a cyberattack, several elements need to be established before the team can assess the risks or take action against the threat (Andrade & Yoo, 2019). For analysts to understand how to manage cyber incidents as a team, they need to have extensive expertise and situational awareness of the cyberspace of the organization, our next mediator in the framework (Figure 4.1). When cybersecurity analysts have more situational knowledge, they can analyze large amounts of information and find patterns with less cognitive effort and more collaboration with other team members. Such situational awareness can lead to the better detection of cyberthreats and more effective teamwork when working in this context.

Cybersecurity teams are unique in that they operate under a time-sensitive and volatile environment, where multiple cyberattacks can happen concurrently. The various possibilities in how the attack will affect the organization or the breadth of data that was (or has the potential to be) breached will determine how individuals and teams react and make decisions. Cybersecurity teams can use the information identified by various team members to create an SMM of potential threats and information that would be useful in dealing with or preventing a cybersecurity attack. Team situational awareness involves three main stages, namely, perception, comprehension, and projection, respectively (Endsley, 1995; Endsley et al., 2003). Team members need to first collectively identify and *perceive* a threat quickly and determine where to focus attention if the information is perceived as a threat. When team members share a collective understanding of the threat and how to proceed, they can further comprehend the problem through interactions with team members.

Comprehension, the second stage of situation awareness, describes how a team will integrate their knowledge and experience using the additional information and resources available from other team members (Endsley, 2000). Members first address the cyberthreat through their own mental model of knowledge and expertise to then refine those thoughts further based on the additional data sources or information brought up by other team members. Importantly, these members may look at the threat from a different perspective, and may provide information that overlaps with the current team members' conceptualization (Cooke et al., 2013). Cybersecurity team members will need to engage in teamwork processes that will facilitate the relay of information between team members (Pala & Zhuang, 2019). In particular, team members who have varying awareness of the situation can exchange information related to the same threat to eventually diverge on an SMM of how to proceed with the threat. By the end of this phase, team members should have an overview of the threat or problem, and have a plan for addressing it as a team.

Projection describes the last phase of situation awareness, which depicts how threats and responses to incidents can be organized across the community of cyber analysts within the organization (Endsley, 2000). This can be described as both a team and organizational phenomenon where analysts discuss common features of cyberattacks to further refine the team's SMM. When these analysts' SMMs converge, team members can use their situation awareness to better predict future needs of the team to resolve the crisis by refining their own understanding of the situation and environment. In summary, team members will need to have an acute understanding of the relevant information shared, as well as how team members will combine perspectives to resolve the cyberattack. This creates a team's SMM that they can use to understand the problem using the different perspectives of team members and how they would resolve the crisis. A better understanding of how these perspectives were derived will provide a more holistic overview of the environment, so team members can work together to determine the most rigorous path to resolving the threat or attack.

Mutual Trust

Team members with specialized knowledge or expertise can share information that is relevant to other team members in the cybersecurity system. As previously mentioned, when team members share this information, they understand the threat or problem through the team's SMM. The team's SMM is a combination of the information or knowledge of team members that is used to guide actions by team members and team outcomes. For high-reliability teams, as in SOC teams, acting on this information will require mutual trust between team members. Mutual trust can be a mechanism (Figure 4.1) for teams to engage in more knowledge integration, where teammates exchange appropriate information with one another and trust that the team member will responsibly act on that knowledge. Similarly, if team members can trust in each other's expertise, they can also apply that information to their understanding of the problem and can learn from one another to further refine their ideas. In essence, when team members develop mutual trust, they can discuss novel threats in greater depth and with more knowledge to build a more effective SMM that is resilient to cyberattacks.

Trust between team members can also encourage members to disseminate relevant information and collaborate with others team members (Jones & George, 1998). It is through trust that team members can work through relevant information that does not coincide with beliefs of other team members to arrive at a consensus on how to proceed with the threat. Increasingly candid discussions can lead to more discoveries of granular information that is relevant to the team. These discoveries, however, can often only be found through interdependent collaboration, coupled with the belief that team members can share information with each other and it can be followed-up on. When team members trust in each other, they learn to integrate their expertise more effectively, leading to more teamwork behaviors that benefit cybersecurity teams.

Cognitive Bias

Given the amount of information that cybersecurity analysts have to interpret, there has been a large discussion in the literature around how to mitigate cognitive biases in cyber decision-making. The danger with cognitive bias in cybersecurity teams is mainly determining whether an incident is a threat (Lemay & Leblanc, 2018). The *base rate fallacy* describes the cognitive error that occurs when individuals do not consider the relative probability of an event after receiving new evidence in the domain (Axelsson, 2000). The base rate fallacy is important to consider when analysts disseminate between the true cyberthreats and false positives. When new information is considered by the team and is perceived as a threat, team members may not feel included to act on the information. This information, therefore, will detract from threats that have a higher probability of causing harm to the organization. Additionally, bias amplification can occur in team settings where various biases and tendencies hamper individual decision-makers' ability to accurately assess cybersecurity threats and make informed decisions (Galin, 2013).

The base rate fallacy may also disrupt the development of the team's SMM and relevant teamwork processes. When teams succumb to the base rate fallacy, they will be working with information that will not inform other team members about the complexities of genuine cybersecurity risks. This would suggest that teams that pursue false threats may not be able to adapt their SMM with the most relevant information. This can also adversely affect the teamwork processes relevant for cybersecurity teams, mainly coordination and communication, because team members will not develop their skills in relation to a legitimate cybersecurity threat (Floodeen et al., 2013). Moreover, when a team experiences an actual event, they may not have the teamwork skills to overcome the threat because much of the team's efforts were toward a perceived threat. Therefore, the adaption of the team's SMM is imperative for cyber teams, considering cyberattacks are consistently evolving and becoming more difficult to detect. To keep up with the attackers' agenda, cyber teams need to determine which threats are genuine so they can train their teamwork processes and SMM to be more accurate when predicting future attacks, while also handling current critical incidents.

Workload

The last mediator, workload, has been shown to affect team outcomes such as team effectiveness, learning and adaptability, and resilience and well-being (Fatima et al., 2022). Team performance is heavily influenced by the workload demands during cybersecurity incidents. When teams become distracted from other teamwork tasks that are not directly related to the threat, they will experience a decrease in team performance and team satisfaction (Lugo et al., 2021). Knowledge integration between team members is necessary to develop a working SMM, but is often cognitively demanding and requires more time to resolve problems. The nature of cybersecurity requires teams to resolve problems quickly, which creates stress on team members who need to engage in teamwork to resolve the crisis. When the workload caused by teamwork or based within the environment increases, team members are more likely to make mistakes or miss critical information necessary for resolving the crisis. These critical insights that are missed may result in a catastrophic cyber event, but will also reduce the strength of a team's SMM harming the future performance of the team.

Similar literature has shown that workload related to teamwork tasks may also harm team member satisfaction. As a consequence of higher workload, team members may feel a greater sense of fatigue and frustration as they work with team members and with the task (Dykstra & Paul, 2018). Stress, as a response during cybersecurity operations, inhibits team members from making accurate decisions and makes them increasingly susceptible to errors. Stress can also create conflict between team members where relevant information may not be communicated effectively among the team. In total, the complexity and unpredictability of cyberattacks causes stress, which affects teamwork and the ability for members to consider new perspectives from others within the team. Therefore, the team is inhibited in more present decision-making but also in regards to how their SMM develops and adapts to new cyberthreats.

FRAMEWORK OUTPUTS

Outputs, the "results" or "outcomes" of the framework, encompass certain performance elements or team member affect (Mathieu et al., 2008, as cited in Grossman et al., 2017). In this case, the major goals for teams of SOC analysts involve increased team effectiveness, team learning and adaptability, as well as team resilience to future threats. Each evidence-based outcome is detailed in the following sections.

DEVELOPING SOC TEAM EFFECTIVENESS

Team effectiveness can be defined as the caliber of team performance outcomes (Fernández Castillo et al., 2023), or the result of a team's capacity to integrate knowledge, work together, and function as a "well-oiled machine". The boom of team effectiveness research dates back to the 1960s, when McGrath introduced the input-process-output (IPO) model that aimed to discern what factors contributed to team effectiveness and performance (Mathieu et al., 2019). Since its publication, researchers have moved past the IPO framework and embraced the dynamic systems teams are nested in (Mathieu et al., 2019). In doing so, team science now recognizes the complexity of team effectiveness, with a multitude of factors contributing to successful teamwork, such as task complexity, team member attributes, team cohesion, boundary spanning, information sharing, psychological safety, leadership, and more (Mathieu et al., 2019). In this section, we review the key drivers of team effectiveness and discuss how SOC teams can incorporate these factors to promote successful outcomes (Tannenbaum & Salas, 2021).

Effective SOC Teams Have the Right Capabilities

Several factors influence team effectiveness, one being team diversity. Diversity can be conceptualized in two buckets: surface-level diversity (e.g., age, race) and deep-level diversity (e.g., educational background). In general, surface-level diversity has a small negative relationship with team performance (see Phillips et al., 2006), but these effects usually disappear over time. In some contexts, surface-level diversity has been found to produce beneficial effects, but in order to benefit from them, organizations should invest and consider inclusive climates and mitigating discrimination (Jansen & Searle, 2021). However, one finding consistent across the literature is that in order for teamwork to be successful, teams need to have the correct combination of knowledge, skills, abilities, and other attributes (KSAOs; Tannenbaum & Salas, 2021). For example, in order for long-duration space missions to be successful, potential astronauts are assessed on a variety of constructs, from leadership capacity to communication – but also on technical knowledge that pertains to the mission itself (Landon et al., 2018). These findings are translatable and generalizable, and in the cybersecurity realm, while not everyone needs to know how to code (i.e., task-related knowledge); everyone must, to a minimum extent, know how to collaborate for teams to succeed (i.e., teamwork-related knowledge). Therefore, successful teams usually exhibit deep-level diversity, and seek to integrate different viewpoints to complete the team's goal.

Effective SOC Teams Cooperate with Each Other

In order to execute a goal, an effective team must be willing to cooperate with each other. As an emergent state, cooperation grows over time (Tannenbaum & Salas, 2021), and is affected by factors like trust and team cohesion. Meta-analytic evidence has shown that intrateam trust is positively related to team performance (De Jong et al., 2016), and factors such as task interdependence can moderate the relationship (De Jong et al., 2016). This suggests that if teams want to be successful, they have to be on the same page in relation to commitment to the task (Grossman et al., 2022). Team members need to have mutual respect and pride in the team's mission (Grossman et al., 2022).

Teamwork is often mistakenly considered an inherent skill, but in order to successfully work as a unit, team members must coordinate efforts by discussing monitoring behaviors, expectations, and capabilities (Tannenbaum & Salas, 2021). By discussing roles and setting norms, teams are better able to understand how to resolve problems that arise. Moreover, successful teams will engage in boundary spanning behaviors, defined as engaging in relationships and interactions with members extant to the team in order to meet objectives (Marrone et al., 2007). Research has shown that teams that do so are able to reduce team member role overload, which can directly explain variability in team effectiveness (Marrone et al., 2007).

Effective SOC Teams Communicate Appropriately and Are on the Same Page

An overload of communication is occasionally confused with better teamwork outcomes, however, quality is preferred over quantity (Tannenbaum & Salas, 2021). Effective teams do this by sharing information, which has been correlated with better team performance outcomes (Mesmer-Magnus & DeChurch, 2009). More specifically, research has found that there are different phases of information sharing: the structuring phase (i.e., norm-setting), the information-sharing phase, and the decision-making phase (Uitdewilligen & Waller, 2018). The most effective teams spend significant time in the structuring and information sharing phase, which then enables them to act more promptly during decision-making (Uitdewilligen & Waller, 2018). By sharing information and appropriately communicating, teams can develop shared cognitions (Tannenbaum & Salas, 2021). Over time, the development of these shared cognitions, more commonly known as SMMs, allows for more effort, improved routine coordination, and faster adjustments (Tannenbaum & Salas, 2021).

Effective SOC Teams Are Aware of Their Surrounding Conditions

Psychological safety (Edmondson, 1999) is one of the pillars for a successful team. Psychological safety is defined as the ability to voice questions, ideas, or problems in the workplace without the fear that one will be negatively reprimanded (Edmondson, 1999). Team leaders have an important role in fostering psychological safety within their teams – and they must set standards by encouraging people to speak up, thanking members for their diverse perspectives, and creating an atmosphere of respect (Tannenbaum et al., 2023).

Effective SOC Teams Coach Each Other

Getting to a destination without a map is challenging, and maybe even impossible. Leadership is another pillar essential for team effectiveness (Zaccaro et al., 2009). Though a plethora of research has remarked on the benefits of different types of leadership styles (in everything from ethical to transformational leadership), the essence of it remains the same: leaders must connect, integrate, and make a cohesive team (Zaccaro et al., 2009). One way to do this is via team coaching (Tannenbaum & Salas, 2021), originally conceptualized by Hackman and Wageman (2005), who posited that by focusing on the team and guiding members; motivation, interpersonal processes, and team knowledge and learning may be improved (Fernández Castillo & Salas, 2023). However, team coaching does not have to be reserved for the team leader, as all team members can participate in helping members understand tasks, goals, and other teamwork-related capacities.

Increased team effectiveness in this context may involve heightened cybersafety, including making less errors when identifying threats and when attacking or mitigating threats. Team effectiveness is vital to the success of teams generally, and is hence a vital aspect to this framework.

Developing SOC Team Learning and Adaptability

Cybersecurity teams must adapt to the varying threats of cyberattackers that have become more devious and complex as technological capabilities have increased. Oftentimes, cyberattacks occur without warning, requiring teams to assemble resources and team members quickly to prevent further damage of the organization or entity. For teams to adapt to the changing circumstances, they need to engage in different team performance behaviors that can overcome the cyberthreat (Steinke et al., 2015). Similarly, team members can also use the perspectives of other analysts to adapt how they approach preventing a cyberthreat or working together to resolve a cyberattack. For teams to become more adaptable they must learn how their teamwork necessitates better team effectiveness (Kozlowski & Bell, 2007). When teamwork failures occur, team members can reflect on how their interactions contributed to the failure.

Adaptable SOC Teams Debrief to Develop a Shared Understanding (Maynard et al., 2020)

Cybersecurity teams can engage in debriefing after a cyberattack to address how team members interacted efficiently and where teamwork breakdowns occurred. As teams debrief, they discuss recent prior events in order to engage in a similar future event more efficiently (Allen et al., 2018; Scott et al., 2022). Considering the volatility of problems faced by cybersecurity teams, engaging in debriefing will allow them to organize their understanding of how environmental, situational, and teamwork factors contributed to teamwork failures. Debriefing after significant cyberattacks can also help teams better understand how they can utilize the roles and perspectives to adapt their expertise. As a result, team members who have experienced several cyberattacks within an entity will develop a shared understanding through the debriefing process preventing future attacks.

Adaptable SOC Teams Acquire New Information and Learn through Perspective-Taking (Calvard et al., 2023)

The team's shared understanding can be influenced through debriefing, but the mechanism behind the development of the team's shared understanding is learning. As team members work closely to resolve crises, they learn new information, but also how to engage in more effective teamwork, which makes cybersecurity teams more adaptable. When team members interact, the combination of new knowledge allows team members to expand their knowledge and consider a new perspective to the problem (Calvard et al., 2023). Perspective-taking helps the analyst become familiar with another component of the cybersecurity system, creating an opportunity for team members to use their knowledge to resolve deficiencies in the system. Taking an innovative approach to cybersecurity will allow for these deficiencies to be identified by internal team members and not because it was found as a weak point or attackers to use to damage the entity. As a result, teams will become more adaptable to the changing cybersecurity domain prior to an attack occurring.

Developing SOC Team Resilience and Well-Being

Finally, the final outcome highlighted in this framework involves team resilience and well-being. Cybersecurity professionals face an array of challenges everyday. Oftentimes, the stress they face is related to a multiplicity of negative work outcomes, such as being less willing to use new technology or adapt to new advances, lower instances of compliance, and more turnover (Singh et al., 2023). Yet, given that many cybersecurity professionals work in teams, this stress is often compounded by teamwork-related issues, such as competing priorities, power differentials, and a lack of experience working together (see Tannenbaum et al., 2023). This stress leads to an increase in collective allostatic load (CAL), the "wear and tear" that groups accumulate over repeated strain (Rosen et al., 2008; Davaslioglu et al., 2021). As stress increases, a team's ability to make decisions is often negatively affected, as cognitive resources begin to overwhelm, making catastrophic errors more likely (Burke et al., 2018). Therefore, in the realm of cybersecurity, a field recognized for constant adversarial threats (Singh et al., 2023), team resilience is of the utmost importance. Defined as a team's capacity to withstand and overcome stressors in a manner that enables sustained performance and well-being (see Alliger et al., 2015, for original definition; Traylor et al., 2023, for the addition of well-being), team resilience has been theorized to increase a team's capacity to improvise, manage conflict, and minimize future stressors (Stoverink et al., 2020). More generally, extant fields have demonstrated the link between resilience, job satisfaction, and psychological strength (Mantas Jiménez et al., 2022), as well as resilience and well-being. In this section, using the current literature on team resilience (e.g., Alliger et al., 2015; Stoverink et al., 2020; Traylor et al., 2023), we overview how teams can minimize, manage, and mend stressors using team-level resources, offering up guiding principles of team resilience.

Minimizing Stress

Resilient SOC teams have shared mental models (Stoverink et al., 2020).

Mental models are defined as structures that allow individuals to interact with their environment (Mathieu et al., 2000). SMMs are created by interacting with one another, and are capable of predicting team performance (Mathieu et al., 2000). Team member ability to converge on the knowledge of the task at hand, share and disseminate information with one another is what creates a successful SMM. Moreover, SMMs are created via the establishment of norms and role-defining (Tannenbaum & Salas, 2021). For example, by meeting before an important task, team members can define who is going to do what, and who they can go to if any questions arise. By doing this, teams become capable of understanding and anticipating risks before they arise, important for the creation of team resilience (Alliger et al., 2015).

Resilient SOC teams are psychologically safe (Stoverink et al., 2020).

Given the high-stakes environment of cybersecurity (Singh et al., 2023), psychological safety might be difficult to foster. Psychological safety is essential for effective team functioning (Tannenbaum & Salas, 2021), and in order to successfully overcome adversity, teams need to be able to communicate concerns with each other without feeling threatened (Stoverink et al., 2020). In order to create psychological safety, teams should seek to encourage all team members (regardless of tenure) to speak up, giving everyone an opportunity to voice questions and concerns (Tannenbaum et al., 2023). Moreover, teams should center feedback on the task (e.g., "debugging code is easier by…") and not on individual characteristics (e.g., "How do you not know how to debug code?") in order to avoid eliciting negative responses from team members (see Hughes et al., 2016). Team leaders can further ease these processes by being open to feedback themselves and embracing vulnerability (Tannenbaum et al., 2023).

Managing Stress

Resilient SOC teams coordinate effectively (Traylor et al., 2023).

Teamwork is often put on the backburner, with many failing to understand the importance of coming together in order to manage a problem. However, research on team cohesion and performance has shown that teams that are capable of focusing and committing to the task at hand effectively outperform those that do not (Grossman et al., 2022). Therefore, in order to increase team resilience, teams should seek to assess challenges and communicate before the task *and* during the task. For example, under a cyberattack, it might seem time consuming to ask a team member to help regarding something they have expertise in. However, doing so might end up saving the team time and resources when resolving the challenge. After all, resilient teams trust each other and are capable of understanding where team member strengths and weaknesses lie (Traylor et al., 2023).

Resilient SOC teams coach other (Traylor et al., 2023).

Team coaching is a team development intervention (TDI; Lacerenza et al., 2018) that seeks to guide teams to be more effective (Hackman & Wageman, 2005). In extant fields, it has been shown to increase team member effort, improve the team's interpersonal processes, and increase the team's knowledge and learning (Fernández Castillo & Salas, 2023). Team coaching can be provided by an external coach, an internal team leader, or in a shared-leadership style, where all team members coach each other. Regardless of the type of coaching, team members should seek to back each other up and provide constructive and positive feedback when needed. By offering guidance to and from each other, teams create trust and help fuel effective team dynamics (Traylor et al., 2023).

Mending Stress

Resilient SOC teams debrief (Alliger et al., 2015).

Debriefs can yield individual and team performance improvements by up to 25% (Tannenbaum & Cerasoli, 2013). In order to become more resilient over time, teams must develop the capacity to look at prior mistakes and prior situations, evaluate them, and discuss what went wrong and what could have been done better. By doing so, teams develop their capacity to deal with incoming stressors, addressing stress more effectively over time and practice (Alliger et al., 2015).

Resilient SOC teams show appreciation for all team members (Alliger et al., 2015).

After a stressful experience, expressing appreciation for team members' efforts can be a successful way to mend (Alliger et al., 2015). Research has found that people consistently feel as though their contributions are undervalued, which can lead to declined team functioning. For this reason, team leaders and members can express appreciation for their team doing simple things such as vocalizing a simple compliment or every once in a while, highlighting one team member's efforts (Tannenbaum et al., 2023). By doing so, team members grow in their social resources and can better attack a challenge in the future.

Resilience is crucial for teams working in extreme environments (Maynard et al., 2020), and well-being is important for SOC-retention and mental health. Resilience and well-being in this context may include better job satisfaction, individual mental health outcomes, employee retention, and an increased ability of the team to continue in spite of continuous threats and challenges. Resilient teams will have the capacity to take on such challenges and learn from their previous experiences.

FUTURE DIRECTIONS FOR RESEARCH IN CYBERSECURITY

This chapter provides an insight into team processes within SOCs. However, there is currently limited research that explores these processes empirically. As such, robust empirical research is needed to delve deeper into the intricacies of team processes in SOCs that can effectively navigate cybersecurity threats. Additionally, in light of the rapid technological innovations in today's world, cybersecurity personnel must be prepared to adapt and continually update their knowledge regarding new potential threats associated with such novel and unknown technology. Concurrently, technological developments are also providing tools to fortify organizational defenses and cybersecurity initiatives. The development of artificial intelligence (AI) has been marked by ongoing innovations, offering promising avenues for exploration and development. Within the realms of team and psychological science, compelling opportunities have been revealed for the utilization of autonomous agents to assist humans in diverse workplace activities.

Importantly, the team science literature has demonstrated the ability of autonomous agents to work alongside humans, a phenomenon termed *human-autonomy teaming* (HAT; Lyons et al., 2021; O'Neill et al., 2022). The demonstrated efficacy of HAT structures introduces boundless prospects to enhance and facilitate transformative changes for SOC teams. These HAT dynamics have the capacity to have critical implications for key teamwork outcomes discussed in this chapter, and future research should focus their attention on these new teamwork structures and their capabilities.

CONCLUSION

Ultimately, SOC teams must be prepared to face the consistent and ever-evolving threats presented to them due to the nature of their occupation. Cybersecurity professionals must collectively navigate daily challenges, drawing from past experiences to make informed decisions and adapting to address future challenges head-on and building team resilience. Job satisfaction and retention issues have emerged as major concerns within the cybersecurity workforce, particularly in SOCs (Adetoye & Fong, 2023). Utilizing the proposed framework (Figure 4.1) detailed in this chapter, SOC teams can optimize their team dynamics, yielding improved outcomes such as team learning, adaptability, resilience, well-being, and ultimately, team effectiveness.

ACKNOWLEDGMENT

This work was partially supported by the US Army Research Institute (ARI) for the Behavioral and Social Sciences and was accomplished under Cooperative Agreement Number W911NF-19-2-0173. This work was also partially supported by NSF award #2205454.

REFERENCES

Adetoye, B., & Fong, R. C. (2023). Building a Resilient Cybersecurity Workforce: A Multidisciplinary Solution to the Problem of High Turnover of Cybersecurity Analysts. In H. Jahankhani (Ed.), *Cybersecurity in the Age of Smart Societies* (pp. 61–87). Springer International Publishing. https://doi.org/10.1007/978-3-031-20160-8_5

Agyepong, E., Cherdantseva, Y., Reinecke, P., & Burnap, P. (2020). Challenges and performance metrics for security operations center analysts: A systematic review. *Journal of Cyber Security Technology*, *4*(3), 125–152. https://doi.org/10.1080/23742917.2019.1698178

Allen, J. A., Reiter-Palmon, R., Crowe, J., & Scott, C. (2018). Debriefs: Teams learning from doing in context. *American Psychologist*, *73*(4), 504.

Alliger, G. M., Cerasoli, C. P., Tannenbaum, S. I., & Vessey, W. B. (2015). Team resilience. *Organizational Dynamics*, *44*(3), 176–184. https://doi.org/10.1016/j.orgdyn.2015.05.003

Andrade, R. O., & Yoo, S. G. (2019). A comprehensive study of the use of LoRa in the development of smart cities. *Applied Sciences*, *9*(22), 4753.

Antonacci, A. C., Dechario, S. P., Antonacci, C., Husk, G., Patel, V., Nicastro, J., … & Jarrett, M. (2021). Cognitive bias impact on management of postoperative complications, medical error, and standard of care. *Journal of Surgical Research*, *258*, 47–53.

Axelsson, S. (2000). The base-rate fallacy and the difficulty of intrusion detection. *ACM Transactions on Information and System Security*, *3*(3), 186–205.

Bammer, G. (2018). Strengthening community operational research through exchange of tools and strategic alliances. *European Journal of Operational Research*, *268*(3), 1168–1177. https://doi.org/10.1016/j.ejor.2017.09.041

Bell, S. T., Brown, S. G., Colaneri, A., & Outland, N. (2018). Team composition and the ABCs of teamwork. *American Psychologist*, *73*(4), 349–362. https://doi.org/10.1037/amp0000305

Bisbey, T. M., Traylor, A. M., & Salas, E. (2021). *Transforming Teams of Experts into Expert Teams: Eight Principles of Expert Team Performance*. *4*(2).

Bowers, C. A., Braun, C. C., & Morgan Jr, B. B. (1997). Team workload: Its meaning and measurement. In M. T. Brannick, E. Salas, & C. Prince (Eds.), *Team Performance Assessment and Measurement* (pp. 97–120). Psychology Press.

Burke, C. S., Priest, H. A., Salas, E., Sims, D., & Mayer, K. (2018). Stress and teams: How stress affects decision making at the team level. In J. Szalma & P. A. Hancock (eds.), *Performance under stress* (pp. 197–224). CRC Press.

Calvard, T., Cherlin, E., Brewster, A., & Curry, L. (2023). Building perspective-taking as an organizational capability: A change intervention in a health care setting. *Journal of Management Inquiry*, *32*(1), 35–49.

Cauwelier, P., Ribiere, V. M., & Bennet, A. (2019). The influence of team psychological safety on team knowledge creation: A study with French and American engineering teams. *Journal of Knowledge Management*, *23*(6), 1157–1175.

Cooke, N. J., Champion, M., Rajivan, P., & Jariwala, S. (2013). Cyber situation awareness and teamwork. *EAI Endorsed Transactions on Security and Safety*, *1*(2).

Davaslioglu, K., Pokorny, B., Sagduyu, Y. E., Molintas, H., Soltani, S., Grossman, R., & Bowers, C. (2021). Collective allostatic load measures for teams. *Structural Health Monitoring*, *20*(3), 768–781. https://doi.org/10.1177/1475921719884864

De Jong, B. A., Dirks, K. T., & Gillespie, N. (2016). Trust and team performance: A meta-analysis of main effects, moderators, and covariates. *Journal of Applied Psychology*, *101*(8), 1134–1150. https://doi.org/10.1037/apl0000110

Dietz, A. S., Driskell, J. E., Sierra, M. J., Weaver, S. J., Driskell, T., & Salas, E. (2017). Teamwork under Stress. In E. Salas, R. Rico, & J. Passmore (Eds.), *The Wiley Blackwell*

Handbook of the Psychology of Team Working and Collaborative Processes (pp. 297–315). John Wiley & Sons, Ltd. https://doi.org/10.1002/9781118909997.ch13

Driskell, J. E., & Salas, E. (1992). Collective behavior and team performance. *Human Factors, 34*(3), 277–288.

Driskell, T., Salas, E., & Driskell, J. E. (2018). Teams in extreme environments: Alterations in team development and teamwork. *Human Resource Management Review, 28*(4), 434–449.

Dykstra, J., & Paul, C. L. (2018). Cyber operations stress survey (COSS): Studying fatigue, frustration, and cognitive workload in cybersecurity operations. *11th USENIX Workshop on Cyber Security Experimentation and Test (CSET 18).*

Edmondson, A. (1999). Psychological safety and learning behavior in work teams. *Administrative Science Quarterly, 44*, 350–383. https://doi.org/10.2307/2666999

Edmondson, A. C., & Lei, Z. (2014). Psychological safety: The history, renaissance, and future of an interpersonal construct. *Annual Review of Organizational Psychology and Organizational Behavior, 1*(1), 23–43. https://doi.org/10.1146/annurev-orgpsych-031413-091305

Eldor, L. (2020). How collective engagement creates competitive advantage for organizations: A business-level model of shared vision, competitive intensity, and service performance. *Journal of Management Studies, 57*(2), 177–209. https://doi.org/10.1111/joms.12438

Endsley, M. R. (1995). Measurement of situation awareness in dynamic systems. *Human Factors, 37*(1), 65–84.

Endsley, M. R. (2000). *Situation awareness analysis and measurement, chapter theoretical underpinnings of situation awareness: A critical review.* Mahwah, NJ: Lawrence Erlbaum Associates.

Endsley, M. R., Bolté, B., & Jones, D. G. (2003). *Designing for Situation Awareness: An Approach to User-Centered Design.* CRC Press.

Fatima, T., Bilal, A. R., Waqas, M., & Imran, M. K. (2022). Keeping up with this workload is difficult: the ramifications of work overload on career resilience. *Kybernetes.*

Fernández Castillo, G., Khalid, M., & Salas, E. (2023). Team Performance. *Oxford Bibliographies in Management.* R. Griffin (Ed.), New York: Oxford University Press. https://doi.org/10.1093/OBO/9780199846740-0220

Fernández Castillo, G., & Salas, E. (2023). Can team coaching provide healthcare the remedy it needs? *Journal of Interdisciplinary Care*, 1–11. https://doi.org/10.1080/13561820.2023.2285030

Fiore, S. M., Rosen, M. A., Smith-Jentsch, K. A., Salas, E., Letsky, M., & Warner, N. (2010). Toward an understanding of macrocognition in teams: Predicting processes in complex collaborative contexts. *Human Factors: The Journal of the Human Factors and Ergonomics Society, 52*(2), 203–224. https://doi.org/10.1177/0018720810369807

Floodeen, R., Haller, J., & Tjaden, B. (2013). Identifying a shared mental model among incident responders. *2013 Seventh International Conference on IT Security Incident Management and IT Forensics* (pp. 15–25). IEEE.

Galin, A. (2013). Endowment effect in negotiations: Group versus individual decision-making. *Theory and Decision, 75*, 389–401.

Gore, J., Ward, P., Conway, G. E., Ormerod, T. C., Wong, B. L. W., & Stanton, N. A. (2018). Naturalistic decision making: Navigating uncertainty in complex sociotechnical work. *Cognition, Technology & Work, 20*(4), 521–527. https://doi.org/10.1007/s10111-018-0531-x

Grossman, R., Friedman, S. B., & Kalra, S. (2017). Teamwork Processes and Emergent States. In E. Salas, Rico, R. & Passmore, J. (Eds.), *The Wiley Blackwell Handbook of the Psychology of Team Working and Collaborative Processes* (1st ed., pp. 243–269). Wiley. https://doi.org/10.1002/9781118909997.ch11

Grossman, R., Nolan, K., Rosch, Z., Mazer, D., & Salas, E. (2022). The team cohesion-performance relationship: A meta-analysis exploring measurement approaches and the changing team landscape. *Organizational Psychology Review, 12*(2), 181–238. https://doi.org/10.1177/20413866211041157

Hackman, J. R. (1987). The design of work teams. In J.W. Lorsch (Ed.), *Handbook of Organizational Behavior*, pp. 315–42. Englewood Cliffs, NJ: Prentice-Hall

Hackman, J. R., & Wageman, R. (2005). A theory of team coaching. *The Academy of Management Review, 30*, 269–287. https://doi.org/10.5465/AMR.2005.16387885

Hagemann, V., Watermann, L., Klonek, F., & Heinicke, C. (2023). Communication quality affects performance of astronauts and support teams through increased workload: Insights from the AMADEE-20 analog mars mission. *Acta Astronautica, 210*, 162–175.

Hughes, A. M., Gregory, M. E., Joseph, D. L., Sonesh, S. C., Marlow, S. L., & Lacerenza, C. N. et al (2016). Saving lives: A meta-analysis of team training in healthcare. *Journal of Applied Psychology, 101*, 1266–1304. https://doi.org/10.1037/apl0000120

Jansen, A. E., & Searle, B. J. (2021). Diverse effects of team diversity: A review and framework of surface and deep-level diversity. *Personnel Review, 50*(9), 1838–1853.

Jones, G. R., & George, J. M. (1998). The experience and evolution of trust: Implications for cooperation and teamwork. *Academy of Management Review, 23*(3), 531–546.

Kim, S., Lee, H., & Connerton, T. P. (2020). How psychological safety affects team performance: Mediating role of efficacy and learning behavior. *Frontiers in Psychology, 11*, 1581–1581. https://doi.org/10.3389/fpsyg.2020.01581

Klasmeier, K. N., & Rowold, J. (2020). A multilevel investigation of predictors and outcomes of shared leadership. *Journal of Organizational Behavior, 41*(9), 915–930.

Klein, G. (2008). Naturalistic decision making. *Human Factors: The Journal of the Human Factors and Ergonomics Society, 50*(3), 456–460. https://doi.org/10.1518/001872008X288385

Klein, G., & Crandall, B. W. (2018). The Role of Mental Simulation in Problem Solving and Decision Making. In P. Hancock, Flach, J., Caird, J. & Vicente, K. (Eds.), *Local Applications of the Ecological Approach to Human-Machine Systems* (1st ed., pp. 324–358). CRC Press. https://doi.org/10.1201/9780203748749-11

Kokulu, F. B., Soneji, A., Bao, T., Shoshitaishvili, Y., Zhao, Z., Doupé, A., & Ahn, G.-J. (2019). Matched and Mismatched SOCs: A Qualitative Study on Security Operations Center Issues. *Proceedings of the 2019 ACM SIGSAC Conference on Computer and Communications Security*, 1955–1970. https://doi.org/10.1145/3319535.3354239

Kozlowski, S. W., & Bell, B. S. (2007). Team learning, development, and adaptation. In V. Sessa & M. London (Eds.), *Work group learning* (pp. 39–68). Psychology Press.

Lacerenza, C. N., Marlow, S. L., Tannenbaum, S. I., & Salas, E. (2018). Team development interventions: Evidence-based approaches for improving teamwork. *The American Psychologist 73*, 517–531. https://doi.org/10.1037/amp0000295

Landon, L. B., Slack, K. J., & Barrett, J. D. (2018). Teamwork and collaboration in long-duration space missions: Going to extremes. *American Psychologist, 73*(4), 563–575. https://doi.org/10.1037/amp0000260

Lemay, A., & Leblanc, S. (2018). Cognitive biases in cyber decision-making. In *Proceedings of the 13th International Conference on Cyber Warfare and Security* (p. 395).

Lipshitz, R., & Shaul, O. B. (1997). Schemata and Mental Models in Recognition-Primed Decision Making. In *Naturalistic Decision Making* (pp. 293–303). Lawrence Erlbaum Associates, Inc. https://doi.org/10.1518/001872008X288385

Lugo, R. G., Ask, T. F., Sütterlin, S., & Knox, B. J. (2021). The influence of team workload demands during a cyber defense exercise on team performance. *HCI International 2021-Late Breaking Posters: 23rd HCI International Conference, HCII 2021, Virtual Event, July 24–29, 2021, Proceedings, Part II 23* (pp. 545–549). Springer International Publishing.

Lyons, J. B., Sycara, K., Lewis, M., & Capiola, A. (2021). Human-autonomy teaming: Definitions, debates, and directions. *Frontiers in Psychology, 12*, 589585.

Mantas Jiménez, S., Lluch Canut, Ma. T., Roldán Merino, J. F., Reig-García, G., & Juvinyà Canal, D. (2022). Resilience and job satisfaction among out-of-hospital emergency medical service professionals: A cross-sectional multi-centric study. *Journal of Nursing Management, 30*(6), 2084–2092. https://doi.org/10.1111/jonm.13645

Marks, M. A., Mathieu, J. E., & Zaccaro, S. J. (2001). A temporally based framework and taxonomy of team processes. *The Academy of Management Review, 26*(3), 356–376. https://doi.org/10.2307/259182

Marrone, J. A., Tesluk, P. E., & Carson, J. B. (2007). A multilevel investigation of antecedents and consequences of team member boundary-spanning behavior. *Academy of Management Journal, 50*(6), 1423–1439. https://doi.org/10.5465/amj.2007.28225967

Mathieu, J. E., Gallagher, P. T., Domingo, M. A., & Klock, E. A. (2019). Embracing complexity: Reviewing the past decade of team effectiveness research. *Annual Review of Organizational Psychology and Organizational Behavior, 6*, 17–46. https://doi.org/10.1146/annurev-orgpsych-012218-015106

Mathieu, J. E., Heffner, T. S., Goodwin, G. F., Salas, E., & Cannon-Bowers, J. A. (2000). The influence of shared mental models on team process and performance. *Journal of Applied Psychology, 85*(2), 273–283. https://doi.org/10.1037/0021-9010.85.2.273

Mathieu, J., Maynard, M. T., Rapp, T., & Gilson, L. (2008). Team effectiveness 1997-2007: A review of recent advancements and a glimpse into the future. *Journal of Management, 34*(3), 410–476.

Maynard, M. T., Kennedy, D. M., Tannenbaum, S. I., Mathieu, J. E., & Levy, J. (2020). Team adaptation and resilience: Where the literature currently stands and how it applies to long-duration isolated, confined, and extreme contexts. *Psychology and Human Performance in Space Programs*, 151–169.

McGrath, J. E. (1964). *Social Psychology: A Brief Introduction*. New York: Holt, Rinehart & Winston.

Mesmer-Magnus, J. R., & DeChurch, L. A. (2009). Information sharing and team performance: A meta-analysis. *Journal of Applied Psychology, 94*(2), 535–546. https://doi.org/10.1037/a0013773

Mohammed, S., Ferzandi, L., & Hamilton, K. (2010). Metaphor no more: A 15-year review of the team mental model construct. *Journal of Management, 36*(4), 876–910. https://doi.org/10.1177/0149206309356804

Müller, R., & Antoni, C. H. (2022). Effects of ICT shared mental models on team processes and outcomes. *Small Group Research, 53*(2), 307–335.

Newman, A., Donohue, R., & Eva, N. (2017). Psychological safety: A systematic review of the literature. *Human Resource Management Review, 27*(3), 521–535. https://doi.org/10.1016/j.hrmr.2017.01.001

O'Neill, T., McNeese, N., Barron, A., & Schelble, B. (2022). Human–autonomy teaming: A review and analysis of the empirical literature. *Human Factors, 64*(5), 904–938.

Ou, A. Y., Waldman, D. A., & Peterson, S. J. (2018). Do humble CEOs matter? An examination of CEO humility and firm outcomes. *Journal of Management, 44*(3), 1147–1173. https://doi.org/10.1177/0149206315604187

Pala, A., & Zhuang, J. (2019). Information sharing in cybersecurity: A review. *Decision Analysis, 16*(3), 172–196. https://doi.org/10.1287/deca.2018.0387

Paoletti, J., Kilcullen, M. P., & Salas, E. (2020). Teamwork in Space Exploration. In *Psychology and Human Performance in Space Programs* (pp. 195–216). CRC Press.

Paul, C. L., & Dykstra, J. (2017). Understanding operator fatigue, frustration, and cognitive workload in tactical cybersecurity operations. *Journal of Information Warfare, 16*(2), 1–11.

Phillips, K. W., Northcraft, G. B., & Neale, M. A. (2006). Surface-level diversity and decision-making in groups: When does deep-level similarity help? *Group Processes & Intergroup Relations*, *9*(4), 467–482. https://doi.org/10.1177/1368430206067557

Rosen, M. A., Salas, E., Wilson, K. A., King, H. B., Salisbury, M., Augenstein, J. S., … & Birnbach, D. J. (2008). Measuring team performance in simulation-based training: Adopting best practices for healthcare. *Simulation in Healthcare*, *3*(1), 33–41. https://doi.org/10.1097/SIH.0b013e3181626276

Ross, K. G., Lussier, J. W., & Klein, G. (2004). From the Recognition Primed Decision Model to Training. In T. Betsch & S. Haberstroh (Eds.), *The Routines of Decision Making* (pp. 327–341). Psychology Press.

Salas, E., Cannon-Bowers, J. A., & Johnston, J. H. (2014). How Can You Turn a Team of Experts Into an Expert Team?: Emerging Training Strategies. In C. E. Zsambok & Klein, G. (Eds.), *Naturalistic Decision Making*. Psychology Press. *It is therefore especially necessary for these teams to establish and utilize a robust team framework to promote overall team effectiveness, among other critical teamwork outcomes (Salas et al., 2014)*"

Salas, E., Rosen, M. A., & DiazGranados, D. (2010). Expertise-based intuition and decision making in organizations. *Journal of Management*, *36*(4), 941–973. https://doi.org/10.1177/0149206309350084

Santos, C. M., & Passos, A. M. (2013). Team mental models, relationship conflict and effectiveness over time. *Team Performance Management*, *19*(7/8), 363–385.

Santos, C. M., Uitdewilligen, S., & Passos, A. M. (2015). Why is your team more creative than mine? The influence of shared mental models on intra-group conflict, team creativity and effectiveness. *Creativity and Innovation Management*, *24*(4), 645–658.

Scott, Z., O'Curry, S., & Mastroyannopoulou, K. (2022). The impact and experience of debriefing for clinical staff following traumatic events in clinical settings: A systematic review. *Journal of Traumatic Stress*, *35*(1), 278–287.

Simonson, R. J., Keebler, J. R., Lessmiller, M., Richards, T., & Lee, J. C. (2020, December). Cybersecurity teamwork: A review of current practices and suggested improvements. In *Proceedings of the Human Factors and Ergonomics Society Annual Meeting* (Vol. *64*, No. 1, pp. 451–455). Sage CA: Los Angeles, CA: SAGE Publications.

Singh, T., Johnston, A. C., D'Arcy, J., & Harms, P. D. (2023). Stress in the cybersecurity profession: A systematic review of related literature and opportunities for future research. *Organizational Cybersecurity Journal: Practice, Process and People*, *3*(2), 100–126. https://doi.org/10.1108/OCJ-06-2022-0012

Sinlapanuntakul, P., Fausett, C. M., & Keebler, J. R. (2022, September). Exploring Team Competencies in Cybersecurity. In *Proceedings of the Human Factors and Ergonomics Society Annual Meeting* (Vol. 66, No. 1, pp. 1110–1114). Los Angeles, CA: SAGE Publications.

Smith-Jentsch, K., Cannon-Bowers, J., Tannenbaum, S., & Salas, E. (2008). Guided team self-correction: Impacts on team mental models, behavior, and effectiveness. *Small Group Research*, *30*, 303–327.

Steinke, J., Bolunmez, B., Fletcher, L., Wang, V., Tomassetti, A. J., Repchick, K. M., … & Tetrick, L. E. (2015). Improving cybersecurity incident response team effectiveness using teams-based research. *IEEE Security & Privacy*, *13*(4), 20–29.

Stoverink, A. C., Kirkman, B. L., Mistry, S., & Rosen, B. (2020). Bouncing back together: Toward a theoretical model of work team resilience. *Academy of Management Review*, *45*(2), 395–422. https://doi.org/10.5465/amr.2017.0005

Tannenbaum, S. I., & Cerasoli, C. P. (2013). Do team and individual debriefs enhance performance? A meta-analysis. *Human Factors*, *55*(1), 231–245. https://doi.org/10.1177/0018720812448394

Tannenbaum, S., Fernández Castillo, G., & Salas, E. (2023). How to overcome the nine most common teamwork barriers. *Organizational Dynamics*, *52*(4), 101006. https://doi.org/10.1016/j.orgdyn.2023.101006

Tannenbaum, S. I., & Greilich, P. E. (2023). The debrief imperative: Building teaming competencies and team effectiveness. *BMJ Quality & Safety*, *32*(3), 125–128. https://doi.org/10.1136/bmjqs-2022-015259

Tannenbaum, S. I., & Salas, E. (2021). *Teams That Work: The Seven Drivers of Team Effectiveness*. New York, NY: Oxford University Press.

Traylor, A., Tannenbaum, S., Thomas, E. J., & Salas, E. (2023). Teamwork in Organizations: The Best Teams Learn, Adapt, and Are Resilient. In C. L. Pierce & Locke, E. A. (Eds.), *Handbook of Evidence-Based Management* (pp. 365–385). Wiley & Sons, Incorporated.

Uitdewilligen, S., & Waller, M. J. (2018). Information sharing and decision-making in multidisciplinary crisis management teams. *Journal of Organizational Behavior*, *39*(6), 731–748. https://doi.org/10.1002/job.2301

Wegner, D. M. (1987). Transactive Memory: A Contemporary Analysis of the Group Mind. In B. Mullen & G. R. Goethals (Eds.), *Theories of Group Behavior* (pp. 185–208). New York, NY: Springer New York.

Wing, L. S. (2005). Leadership in high-performance teams: A model for superior team performance. *Team Performance Management: An International Journal*, *11*(1/2), 4–11.

Zaccaro, S., Heinen, B., & Shuffler, M. (2009). Team Leadership and Team Effectiveness. In E. Salas, Goodwin, G. F. & Burke, C. S. (Eds.), *Team Effectiveness in Complex Organizations: Cross-Disciplinary Perspectives and Approaches*. Routledge.

Zimmerman, C. (2014). Ten Strategies of a World-Class Cybersecurity Operations Center. The MITRE Corporation. https://duikt.edu.ua/uploads/l_1717_91042607.pdf

5 Insights and Recommendations for Improving Handoff Processes in Cybersecurity Operations Centers

Liam Brennan, Crystal M. Fausett, and Elizabeth H. Lazzara

INTRODUCTION

Within the ever evolving and integrating technoscape that cybersecurity professionals find themselves operating, the utility of effective and efficient teams cannot be understated. Gone are the days when a lone employee could effectively manage an organization's technological infrastructure, in addition to monitoring for, and responding to cyber threats. With advances in the systems they are tasked with safekeeping, so too are advances in the integration of cybersecurity professionals and their teams required. This need for robust teamwork is particularly acute in the critical realm of cybersecurity operations. The impacts of a cybersecurity incident can extend far beyond the technical realm in which it began, leading to economic consequences (Furnell et al., 2020), legal issues (Shackelford, 2014), and disruptions to psychological well-being (Alashwali et al., 2024). As organizations increasingly depend on the internet to conduct their operations, the frequency and severity of cyberattacks also continue to escalate. In response, cybersecurity operations centers (CSOCs, or simply SOCs), play a crucial role in monitoring, detecting, and mitigating cyber threats (Crowley, 2024). SOCs play a critical role in cybersecurity, with most medium and large organizations maintaining one in some form. For organizations that lack the resources to establish their own SOC, third-party service providers offer essential SOC functions to ensure security and threat management (Miloslavskaya, 2016). The significance of SOCs continues to grow as cybersecurity becomes essential across various sectors, including businesses, government agencies, nonprofits, and academic institutions (Knerler et al., 2022).

However, little research has been conducted on the handoff processes that are integral to SOC operations (Fausett & Keebler, 2022). Effective handoffs (transferring information, tasks, and responsibilities between personnel) are likely to be

DOI: 10.1201/9781003413707-6

crucial in ensuring continuous and coordinated responses to cybersecurity incidents (Crowley, 2021). Unsuccessful handoffs present challenges such as missing information, incorrect assumptions, and dropped tasks, particularly in high-pressure SOC environments (Kaplan, 2023). Understanding SOC teamwork from a team cognition perspective is essential for supporting their high-pressure collaborative work (Hámornik & Krasznay, 2018). We argue that human factors, industrial and organizational psychology, cognitive, safety, and team science offer a valuable lens to better understand how to overcome handoff challenges. Drawing on research from other safety-critical industries, such as healthcare, nuclear power, aviation, and oil and gas, we illustrate the importance of handoff operations and explore how similar handoff strategies could be adopted by SOCs. SOCs rely on seamless operations, yet handoffs between analysts often present significant challenges. This chapter aims to bridge the gap between the critical need for effective handoffs and the limited research available.

The SOC serves as an important component of an organization's cybersecurity efforts, where teams of analysts monitor, detect, and respond to security threats. Within this environment, the handoff process represents a potentially vulnerable point in operations that has received limited attention in research and practice. This chapter examines SOC handoffs through multiple perspectives: team science principles, healthcare handoff models, and selected cybersecurity case examples. While no single approach can address all handoff challenges, we aim to provide insights that may help security professionals consider their current practices. For security practitioners, this chapter offers potential frameworks adapted from other domains with more established handoff protocols. For researchers, we identify some knowledge gaps that might benefit from further study. For organizational leaders, we discuss how handoff quality may impact operational effectiveness.

By connecting knowledge from team science and healthcare with the specific context of cybersecurity operations, this chapter attempts to contribute to an emerging conversation about this aspect of SOC functioning. The approaches and considerations presented here should be viewed as starting points rather than definitive solutions, as the field continues to develop more specialized knowledge about effective handoffs in cybersecurity environments. We begin by defining key terms: SOCs, teamwork, and handoffs. Next, we examine the dearth of research on this topic and the consequences of handoff failures in various industries. Building on this foundation, we then present considerations specific to SOC handoffs, discuss recommended handoff practices, and examine a handoff mnemonic introduced in a 2021 Security Boulevard blog post (Crowley, 2021) followed by an example verbal handoff and discussion. This chapter concludes with a research agenda.

WHAT IS A SOC?

A SOC can be defined as "a team primarily composed of security analysts organized to detect, analyze, respond to, report on, and prevent cybersecurity incidents" (Knerler et al., 2022). This definition aligns with established incident response frameworks, such as those outlined by the National Institute of Standards and Technology

(NIST SP 800-61; Cichonski et al., 2012). There are actually many terms that can be used to describe a team of people, interdependently working together with a shared goal of performing the functions described above. Knerler et al. (2022) explained that some of these terms include:

- Computer Security Incident Response Team (CSIRT)
- Computer Incident Response Team (CIRT)
- Computer Incident Response Center (or Capability) (CIRC)
- Computer Security Incident Response Center (or Capability) (CSIRC)
- Computer Emergency Response Team (CERT)

The various terms used, such as CSIRT, CIRT, and CERT, reflect the evolving nature of cybersecurity operations. Industry reports from organizations such as the SANS Institute highlight the diverse terminology used within the field (Crowley, 2024). Complicating matters further is the interchangeability of these words with other synonyms (e.g., computer in any of the above names could be replaced with cybersecurity; incident could be exchanged for emergency). As of the writing of this chapter, the term *cybersecurity operations center* appears to be the most commonly used term in the reviewed literature, so this is the term that this chapter will use. *Physical* security operations centers are also referred to as "SOCs," so this book includes "cybersecurity" in the title to clarify the focus. However, throughout the text, we use the shorter term "SOC" to refer specifically to cybersecurity operations centers.

There are variations in the size, primary services, and kind of support that may be offered by any given SOC (or CSIRT, etc.), and key operations will typically vary as a result of several factors including organizational size, network infrastructure, and threat landscape (Weissman & Jayasumana, 2020). Variations aside, the core functions of any given SOC all boil down to the definition proved above: incident detection, investigation, and response. However, even with clearly defined functions, the success of a SOC relies heavily on the individuals performing those tasks and their ability to work together. Therefore, to fully understand the dynamics of SOC operations, we must first examine the concept of a team.

WHAT IS A TEAM?

Given the critical role of teamwork in high-pressure SOC environments (Hámornik & Krasznay, 2018), especially during handoffs, it's essential to establish a precise understanding of what constitutes a "team." While Merriam-Webster (n.d.) may define a team as "a number of persons associated together in work or activity", within the realm of teams' research, a more restrictive approach is taken where a "true" team is considered a "distinguishable set of two or more individuals who interact dynamically, adaptively, and interdependently; who share common goals or purposes; and who have specific roles or functions to perform" (Salas et al., 1992). In decomposing this definition, the disparities between teams and various forms of collaborative collectives become clearer.

Distinguishable Set

To some extent, the act of declaring individuals as members of a team (i.e., identifying their membership), contributes toward their meeting the requirements of being a "true" team. A collective unit of individuals with differing positions may constitute a "true" team within the context of a shared project, without being from the same departments (Litchfield et al., 2017). As such, these individuals may or may not recognize themselves as being team members, yet their output is attributed to the collective unit and they subsequently meet the criteria of a "true" team. This recognition of membership and/or collective output, either by themselves, other members, or outside individuals, contributes to their being a team and their overall performance (Ashforth & Mael, 1989; Lin et al., 2016).

Two or More Individuals

While seemingly apparent, the consideration that a team requires a minimum of two people (i.e., there cannot be a team of one) but may vary in their maximum size, is worthwhile. Recognizing that partnerships have the potential to fulfill the requirements needed to be a team and vice versa that large groups may not always do so is of note (Kozlowski & Ilgen, 2006). A small SOC whose employees work together and depend on each other may be considered a team whereas a large SOC may be divided into many separate teams that fulfill different objectives (Weiss & Hoegl, 2015). This collection of many separate teams thus becomes a multi-team system, a concept which is further explored in a later chapter.

Interact Dynamically, Adaptively, and Interdependently

Here is where many collective units fail to meet the requirements of being "true" teams. While many combinations of dynamism, adaptivity, and interdependence may be expressed in various forms across the full population of "true" teams, each underscores aspects of their interactions (exchanges of information, resources, support, etc.). Examining these interactions helps reveal the subtle differences between various types of teams.

- Dynamism refers to the degree to which interactions vary. More dynamic interactions may be more diverse and potentially improvisational while less dynamic interactions are more uniform and potentially scripted. These differences may be influenced by the interplay of the team's processes and emergent states (Marks et al., 2001). Processes encompass the methods by which team members interact with each other, their task, and environment whereas emergent states are "products of the team experience" itself (Benishek & Lazzara, 2019).
- Adaptivity refers to the degree to which interactions evolve over time as a result of prior outcomes and performance. More adaptive interactions may greatly differ from the preceding interactions while less adaptive interactions will be rather consistent over time regardless of past results.

- Interdependence refers to the degree which interactions are required to achieve the desired outcome. More interdependent interactions may require greater or more frequent collaboration between team members; while, less interdependent interactions would require less or more infrequent collaboration. This may vary with differences in what is exchanged (e.g., information, resources; Widianto et al., 2024) as well as the sequencing by which the interaction takes place (e.g., pooled, sequential, reciprocal, or team; Thompson, 1967; Van de Ven et al., 1976).

In this sense, a collective of computer support specialists who each independently responds to user service tickets without dynamically or adaptively interacting, wouldn't constitute a team as they each perform and achieve their tasks without any input from one another. This is further reinforced if their performance is assessed based on their individual ability as opposed to the teams' metrics (Saavedra et al., 1993).

Share Common Goals or Purposes

As with the characteristic of team size, the need for a team to have a collective aim may seem self-evident, but this component, too, is not without value (Larson & Schaumann, 1993). Within an organization, departments of individuals may share a name and interact during their work, but the positions those individuals inhabit and the roles they undertake may work to achieve very different outcomes. As with the previous SOC example, a small department devoted to meeting the needs of a business has a broad but ultimately shared goal/purpose. Comparatively, a large SOC within an organization may have a variety of teams each working to meet different cybersecurity needs (threat assessment, employee education, developing applications, managing infrastructure, etc.).

Specific Roles or Functions to Perform

There are few groups of individuals where said members have no specified role or function. Even if all individuals share the same tasking obligations this will constitute a role on every individual's part.

When considering the application of these requirements to modern cybersecurity professionals and their associated SOCs, the positions and roles that coalesce into "true" teams vary greatly. With this and the aforementioned characteristics of teams in mind, the following sections examine the nature and structure of teams as they pertain to cybersecurity professionals.

NATURE AND STRUCTURE OF TEAMS

In considering the nature of teams, we approach from the perspective that various factors may influence how a team operates without changing the fundamental makeup of said team (transition to remote work, relying on shift work, etc.). By comparison, shifts in the structure of a team fundamentally change the team itself

TABLE 5.1
Overview of Natural and Structural Factors of Teams

Factor Type	Factor	Description
Nature	Distribution	*In the same or different places*
	Virtuality	*Using communication technologies to interact*
	Synchronization	*Working at the same or different times*
	Fluidity	*Able or unable to move in and out of the team*
Structure	Skill Differentiation	*Overlapping or specialized in their knowledge, skills, and abilities*
	Authority Differentiation	*Organized in a vertical or horizontal hierarchy*
	Temporal Stability	*Consistent over time and share a history or future*

(cross-training all team members, reducing the formal leadership structure within the team, etc.).

Understanding differences in the nature of teams is valuable and worthy of consideration, especially within the context of cybersecurity where teams that perform comparable functions may operate in wildly different manners. Vice versa, understanding differences in the structure of teams allows knowledge to be cultivated which can be generalized and applied to any team. Though defining exactly when a team's identity changes remains debated (similar to the Ship of Theseus paradox (Plutarch, ca. 75 A.C.E./1994), which questions whether an object remains the same if all its parts are gradually replaced), the fundamental characteristics of teams are essential for meaningful comparisons both within cybersecurity and across other fields.

In this respect, four key continuums of factors exist which speak to the nature of a team including distribution, virtuality, synchronization, and fluidity. Likewise, three factors surrounding the structure of a team including skill differentiation, authority differentiation, and temporal stability are presented. These structural factors are pulled from the three-dimensional team scaling model (3DTSM; Hollenbeck et al., 2012). Table 5.1 provides an overview of these natural and structural factors of teams as well as brief descriptions to clarify the nuance between them.

Distribution

The concept of distribution speaks to whether individual members of a team are co-located (i.e., housed at the same location) or distributed (i.e., housed at different locations) (Stagl et al., 2007). Even among distributed teams, the degree of separation may vary widely between smaller regional organizations and other multinational conglomerates. SOCs are increasingly distributed; organizations with global operations are likely to have multiple SOCs in different regions to allow for 24/7 coverage (Condello, 2020). Commonly referred to as a "follow the sun" model, this has many benefits for SOCs that are vulnerable to security events at any point in time. This model is a type of global workflow in which security events can be handled by and shared between multiple SOCs in different time zones (Burg, 2020). This allows SOC team members to work during their regular hours, and therefore avoid a majority of the issues that arise with night shifts and long hours such as fatigue and burnout, while still providing around the clock cyber network defense. Other drivers of SOC distribution include geographic proximity to assets, talent acquisition, and cost considerations (Wang et al., 2023). Thoughtful handoff processes can maintain communication across distributed teams.

Virtuality

While inherently linked to distribution, which focuses on where individual team members are located, virtuality speaks to the extent to which team members employ communication technologies (audiovisual calling, emails, messaging, etc.) to interact (Zheng et al., 2024). With this, teams may range across this continuum with a combination of team members who are in-person (i.e., on-site), hybrid (i.e., on-site or off-site part of working time), or remote/virtual (i.e., off-site for all work).

Synchronization

Synchronization differs from distribution and virtuality in that it speaks to whether individual members of a team are working synchronously (i.e., at the same time) or asynchronously (i.e., at different times) regardless of where that time may be spent. Individuals who are distributed across a region and in different time-zones may still work synchronously, despite their virtuality, to facilitate cooperation.

Fluidity

By comparison to the prior three factors, fluidity approaches teams as being either intact (i.e., consistent membership/roster) or fluid (i.e., changing membership/roster) but exactly when a team ceases to be fluid or intact is difficult to discern; most teams lie on a spectrum between these points (Driskell et al., 2023). Intact teams reflect strong boundaries between individuals who are or are not members whereas more fluid teams possess weak boundaries with individuals moving across distinctions (i.e., members or non-members) more freely depending on the organization or environment.

Skill Differentiation

This refers to the degree to which SOC team members have overlapping or specialized knowledge, skills, and abilities. A SOC with high skill differentiation consists of highly specialized analysts, threat hunters, incident responders, and forensic experts, each focusing on distinct tasks with minimal overlap (Cassetto, 2022; see Table 5.2). In contrast, a SOC with low skill differentiation would have cross-trained personnel capable of performing multiple roles. Even within the same SOC team, skill differentiation emerges based on prior training, past job roles, and hands-on experience with different tools and threat landscapes. Any characteristic that has the potential to impact team performance, such as education, culture, or past experiences, can contribute to skill differentiation (Hollenbeck et al., 2012). Understanding

TABLE 5.2
Example SOC Roles

Role	Qualifications	Duties
Tier 1 Analyst (Alert Investigator)	System administration, web programming (Python, Ruby, PHP), scripting, security certifications (CISSP, SANS SEC401)	Monitors SIEM alerts, configures security tools, triages alerts to determine real incidents.
Tier 2 Analyst (Incident Responder)	Similar to Tier 1, plus incident response experience, advanced forensics, malware assessment, threat intelligence, ethical hacking certification	Analyzes incidents, correlates threat intel, identifies threats, and leads containment, remediation, and recovery.
Tier 3 Analyst (SME; Threat Hunter)	Similar to Tier 2, plus penetration testing, malware reverse engineering, cross-org data visualization	Conducts vulnerability assessments, penetration tests, and proactive threat hunting. Assists Tier 2 in major incident response.
Tier 4 SOC Manager (Commander)	Similar to Tier 3, plus project management, incident response leadership, strong communication	Leads SOC team, manages security strategy, resources, and major incident response. Key point of contact for security incidents and compliance.
Security Engineer (Support and Infrastructure)	CS, CE, or IA degree with CISSP or equivalent certification	Designs security solutions, develops tools, and supports SOC operations within development or infrastructure teams.

Source: Adapted from Cassetto (2022).
Note: SME, Subject Matter Expert.

skill differentiation is critical for SOC efficiency, as it affects how incidents are escalated, how information is shared, and how well team members can cover for one another in high-pressure situations.

Authority Differentiation

While skill differentiation reflects horizontal specialization, authority differentiation captures the hierarchy of leadership (Zhou et al., 2017). A highly differentiated SOC follows a strict chain of command, with tiered escalation paths where Tier 1 analysts escalate incidents to Tier 2 and Tier 3 experts, who may then consult SOC managers or CISOs for major decisions (Bykowski, 2023). Conversely, a SOC with low authority differentiation may operate with a more collaborative approach, where analysts have greater autonomy in decision-making and incident response. Formal leadership roles, such as SOC managers and CISOs, inherently hold more authority, but informal leadership, such as experienced analysts mentoring new hires or leading investigations, also plays a crucial role in SOC operations. The degree of authority differentiation has the potential to affect SOC team's psychological safety and innovation (Gao et al., 2024).

Temporal Stability

Temporal stability in a SOC reflects the consistency of team membership over time and the degree to which analysts have a shared history or expected future working together (Hollenbeck et al., 2012). Many SOCs experience high turnover due to burnout, stress, and career advancement opportunities, leading to low temporal stability. This rapid personnel change disrupts the development of team cohesion, implicit communication, and trust, all of which are essential for efficient threat detection and response (Hackman & Katz, 2010). While low temporal stability can introduce fresh perspectives and new skills to a team (Kane et al., 2004), it also limits the opportunity for developing shared norms and coordination behaviors (Reagans et al., 2005) and may undermine psychological safety (Koopmann et al., 2016). Given the high burnout rates among cybersecurity professionals reported both anecdotally (Beyond Machines, 2022; Boyer, 2019; Nobles, 2022) and in the literature (Reeves et al., 2023), it is expected that temporal stability would be relatively low in many SOCs. This is further reflected in the average reported tenure of cybersecurity professionals in senior roles, which is only 18 to 24 months (Nominet, 2019). Implementing structured knowledge transfer, mentorship, and well-documented processes can help mitigate the negative effects of frequent personnel changes. This need for effective knowledge transfer brings us to a critical aspect of team performance in SOCs: handoffs.

WHAT IS A HANDOFF?

Handoffs represent one of the most crucial moments where knowledge transfer and team stability are tested. While we've focused on cybersecurity teams thus far, the most comprehensive research on handoffs comes from healthcare settings, which face similar challenges in maintaining continuity despite personnel changes and shift rotations. By examining healthcare handoffs, we can draw valuable parallels to improve SOC practices.

Handoffs are essential in many industries, domains, and organizations (Patterson et al., 2004), but have been most extensively studied in healthcare. In this context, a handoff can be described as the transfer of both responsibility and accountability of patient care from at least one medical provider to another (Sumner et al., 2019). Effective handoffs ensure continuity of care and prevent errors, as they involve the exchange of critical information, such as patient history, treatment plans, and any ongoing concerns. There are many terms that can describe this transfer of responsibility and accountability, including:

- Change out
- Sign out

- Shift report
- Handover
- Pass down
- Sign over
- Transfer
- Protocol

At a basic level, handoffs can be thought of as having two parties, a sender, and a receiver. The sender is typically the outgoing party, the receiver the incoming party who will be assuming responsibility and accountability of care. An effective handoff requires more than just the transfer of information, it demands clear, structured communication to ensure continuity and prevent errors.

Handoff protocols and methods can differ widely in structure and are often tailored to the specific needs of various domains, industries, and organizations. For instance, in healthcare, standardized protocols like SBAR (Situation, Background, Assessment, Recommendation) are commonly used to ensure clarity and reduce the likelihood of miscommunication (Patterson & Wears, 2010). In aviation, a similar concept called a "cockpit handoff" ensures that important information about the flight's status is passed from one pilot to another during shift changes. In cybersecurity, handoffs often take place between SOC analysts during shift changes, where detailed information about ongoing security incidents is exchanged. Much like in aviation or healthcare, SOCs must ensure that handoffs are precise and accurate to prevent cybersecurity incidents from escalating or being overlooked. However, cybersecurity incidents can evolve rapidly and unpredictably, demanding an even more dynamic and flexible handoff system.

Research on handoff processes in other fields can offer valuable insights into how these transitions might be effectively managed within cybersecurity. However, there is a lack of research focused on the transfer of responsibility between cybersecurity professionals. Similar to other aspects of teamwork (Simonson et al., 2020), handoff processes in cybersecurity have not been extensively studied. In contrast, the domains discussed below have undergone significant scientific examination of handoff practices. Like cybersecurity, these fields also face severe consequences when failures occur. Investigations into catastrophic accidents reveal that breakdowns in shift-work information exchange either through inadequate communication or omission of crucial details played a key role in the events leading up to the disaster (Carvalho et al., 2012).

While the structure of handoff protocols may vary, they all serve the same core function: to facilitate the organized transfer of critical information, responsibility, and accountability (Patterson & Wears, 2009; Nilsen et al., 2018). Effects, either positive or negative, regarding changes in process may not be observed until succeeding shifts; it is therefore important to be aware of the system status in the previous shifts, the current status, and to predict future developments (Nilsen et al., 2018). This is where handoffs become key to ensuring safe operations. Challenges arise when handoff processes are unclear, incomplete, or poorly communicated, leading to misunderstandings, missed threats, or errors. In this chapter, handoffs are considered a domain-independent concept, applicable in a wide range of sectors and critical to ensuring the successful transfer of tasks, duties, and information. Successful handoffs play an important role in ensuring seamless transition of work processes and the safety of organizations, especially those dealing with shift work or frequent transfers of care and responsibility.

A DEARTH OF RESEARCH

Maturity of the Field

Handoff research in healthcare has been extensively studied, likely because of the direct impacts on patient safety. In contrast, cybersecurity operations centers are newer, and as a result structured research on SOCs are still emerging. Handoffs in healthcare have gained even more attention since the Institute of Medicine report *To Err Is Human*, which linked poor communication to medical errors and patient deaths (Institute of Medicine (US) Committee on Quality of Health Care in America, 2000).

SOCs, on the other hand, have grown with the rise of cyberattacks, data breaches, as well as regulatory requirements, although SOCs are highly encouraged by financial sectors and healthcare sectors to ensure compliance with privacy regulations and standards (Berezin, 2025; Miller, 2024).

LACK OF FORMALIZED PROCESSES

SOCs are likely to have unique shift change procedures, making it difficult to generalize research. Widely accepted handoff frameworks are still very much lacking in SOCs. SOCs operate differently across industries, companies, and security teams, resulting in a lack of consistent methodology for which to study handoffs.

It is true that these idiosyncrasies exist within healthcare as well. However, there are a few key reasons why healthcare has still managed to develop a deeper body of research:

- Regulatory pressure
- Public visibility
- Interdisciplinary research culture

Healthcare is heavily regulated, through HIPPA and the Joint Commission standards, which have pushed forward standardization efforts across the industry. In contrast, while SOCs in various sectors face different regulatory requirements such as HIPAA in healthcare organizations, financial regulations in banking and financial services, and FERPA in higher education, there remains less uniformity in SOC operations across these sectors compared to healthcare practices. Healthcare is also much more likely to make headlines and suffer lawsuits when things go wrong. Security failures, on the other hand, require a high degree of catastrophe to be released to the public – corporate concerns and NDAs serving to keep information under lock and key. Finally, human factors and medicine have a long, intertwined history in the pursuit of patient safety and quality. Cybersecurity, on the other hand, historically has placed an emphasis on technical solutions (Etschmaier, 2019).

Classified and Proprietary Data

Researchers in the field of cybersecurity are often faced with challenges around collecting data due to the sensitive nature of cyber events, logs, and workflows. It is more difficult to anonymize this data for research purposes, as SOC data involves highly sensitive information about security threats, organizational defenses, and attack vectors, making it difficult to share or study handoff phenomena externally. Companies are reluctant to admit handoff failures in cybersecurity incidents.

Studying cybersecurity incidents as an outsider presents challenges due to the industry's emphasis on confidentiality, integrity, and availability of information. Post-mortems on cybersecurity failures are rarely shared with external researchers, making it difficult to analyze the role of team communication in specific incidents. However, research has broadly identified communication as a significant issue in SOCs. Studies highlight problems such as information overload, cognitive tunneling, and high false alarm rates among security analysts (Champion et al., 2012). SOC teams often struggle with coordination, trust, and information sharing – factors critical to fostering a positive security culture (Ioannou et al., 2019). Effective threat intelligence sharing is further hindered by insufficient skills and motivation related to reporting and communication (Brilingaitė et al., 2022). Additionally, a systematic literature review by Vielberth et al. (2020) noted that while research focuses on human and technological aspects of SOCs, it frequently neglects the processes that bridge these domains, particularly non-technical processes like communication and handoffs. Aside from successfully performing the handoff activity itself, we argue that there should be substantial consideration devoted to the design and preparation of the complex system in which these handoffs occur.

Solutions to this are tricky. Some of those include pushing for a SOC handoff research focus – encouraging human factors/psychology and cybersecurity professionals to collaborate. There is the potential for university and industry partnerships here. Other options are to benchmark against other high reliability organizations, and adapt these lessons to cybersecurity operations centers, which is the intent of this chapter.

Fewer Observable Consequences

Within critical sectors such as healthcare, as well as in the documented incidents presented in our case studies, the deleterious effects of suboptimal information transfer during handoffs manifest as immediate patient mortality. This underscores the imperative for rigorous investigation into effective communication methodologies. While the immediate consequences of deficient handoffs in SOCs may appear less overtly catastrophic, potentially leading to delayed threat detection and protracted incident response times, this should not be misconstrued as an absence of severe repercussions. The WannaCry ransomware outbreak of 2017, which disrupted healthcare services in the United Kingdom, exemplifies the potential for substantial harm, with research indicating a decrease in hospital admissions and appointments (Ghafur et al., 2019). Moreover, the first documented ransomware-associated death, reported at a German hospital in 2020, further substantiates the life-threatening potential of cyberattacks (Kwon et al., 2021).

Interdisciplinary Barriers

While cybersecurity-specific studies on teamwork are limited, it is reasonable to assume that cybersecurity teams could draw insights from research in other domains. However, cybersecurity presents unique challenges that necessitate dedicated investigation. First, cybersecurity teams operate in an inherently reactive environment where threats emerge and evolve rapidly, requiring a level of adaptability and resilience that few other fields demand. Second, the abstract and complex nature of cybersecurity work places significant cognitive demands on professionals, often exceeding those encountered in other disciplines (Zaccaro et al., 2016). Third, cybersecurity is distinctive in that computers serve as both the primary tools and the objects of the work itself, a dynamic that adds another layer of complexity (Zaccaro et al., 2016).

The intersection of cybersecurity and human factors represents an important but understudied research area. While extensive literature exists on team effectiveness in human factors and industrial-organizational psychology, applying these insights to cybersecurity's unique challenges requires specialized investigation. The reactive nature of cybersecurity work, its cognitive demands, and distinctive human-computer interactions create a context that warrants dedicated human-centered research. Addressing these challenges can not only improve teamwork within cybersecurity but also contribute to broader insights on team effectiveness across disciplines. The study of SOC handoffs requires an intersection of cybersecurity and human factors, which is a niche intersection. Progress in this area has been hampered by practical barriers. Human factors researchers typically lack access to cybersecurity environments, while cybersecurity professionals primarily focus on technical rather than human-centric aspects.

Emphasis on Technical Solutions

Cybersecurity research tends to focus on technical solutions (artificial intelligence, automation, and threat detection), rather than human workflows. This means that handoffs, which are primarily about human communication and process efficiency, are not given much attention. Researchers have identified two key skill sets necessary for effective team performance: taskwork skills and teamwork skills (Burke et al., 2003). According to Salas et al. (2007), taskwork skills encompass the competencies required to execute specific job responsibilities, while teamwork skills involve the attitudes and behaviors that facilitate effective team collaboration. Rather than depending on the abilities of a single individual, organizations leverage teams of specialized, interdependent professionals who possess the necessary knowledge, skills, and abilities (KSAs) to successfully perform task-related duties.

Zaccaro et al. (2016) suggested that enhancing cybersecurity performance similarly requires a dual-focus strategy. First, they emphasized the need for improved intrusion detection systems and other automation tools to reduce the cognitive burden on cybersecurity professionals, ultimately enhancing their decision-making capacity. This aspect aligns with strengthening taskwork skills. Equally important is the second approach, which highlights the role of psychosocial factors in cybersecurity. Enhancing team dynamics and management practices can foster better collaboration, coordination, and overall productivity among cybersecurity teams, particularly given the complexity of cyber-related tasks. This perspective aligns with the development of teamwork skills, which, alongside taskwork skills, are essential for high-functioning teams (Salas & Fiore, 2004). Social interaction also plays a significant role in fulfilling job responsibilities, as noted by the NIST in its 2012 report (NIST, 2012).

Unclear Metrics for Success

In medicine, various metrics have been used to determine handoff success. Although there are a variety of difficulties in handoff measurement within healthcare, it is reasonable to assume that less research has been devoted to the creation of measurements for SOC handoff effectiveness. The authors assume that in-house efforts to evaluate and improve SOCs are increasing, and thus academic literature is beginning to catch up by providing metrics and models for assessing SOC performance (Chamkar et al., 2024). Even so, a majority of published research on this topic attends to the technology and processes with little attention to the human-system interactions (Zimmermann & Renaud, 2019). A literature review conducted by Sundaramurthy et al. (2014) discovered a near absence of adequate metrics for SOC analyst performance, let alone metrics to determine handoff quality and success.

HANDOFF FAILURE CASE STUDIES

Effective handoffs are important for preventing missed threats, ensuring continuity in monitoring, and maintaining incident response integrity. SOC analysts need to work in teams and be able to quickly exchange their findings and divide tasks among each other (Zhong et al., 2019). Having a formal handoff mechanism or protocol can help team processes and performance as they prevent information loss (Fausett & Keebler, 2022). Currently the information sharing and task division are done inefficiently which negatively impacts the response and decision-making time (Zhong et al., 2019). Studies from other high-risk domains emphasize that failed handovers can contribute to large-scale accidents, emphasizing the need for skilled team communication and reliable handover systems (Nilsen et al., 2018). The case studies below highlight the importance of handoff processes, and what happens when handoffs go wrong.

Cybersecurity

The rarity of case studies around cybersecurity incidents makes it difficult to directly trace poor handoff communication to cybersecurity incidents (hence the discussion of case studies from a variety of different domains and industries below). Notably, however, there is one incident discussed in Zaccaro's chapter within this textbook that alludes to poor teamwork as a contributing factor to an incident. The 2017 Equifax Data Breach was responsible for the potential disclosure of social security numbers, addresses, and credit histories of 147 million consumers

(Equifax, 2020). Although handoff failure was not directly implicated in Kabanov and Madnicks' (2020) systematic study of the incident, challenges with information exchange were prevalent.

At the heart of the incident was a known vulnerability in Apache Struts 2, a framework used in Equifax's Automated Consumer Interview System (ACIS; Kabanov & Madnick, 2020). On March 8, 2017, US-CERT (United States Computer Emergency Readiness Team) issued an alert warning organizations to patch the flaw, and Equifax's Information Security (IS) team received and disseminated this alert across the company (Diniyatullah & Rindu, 2024). However, a critical breakdown in communication followed: the individual responsible for ACIS was not included in the distribution list, leaving the system unpatched (Kabanov & Madnick, 2020). Additionally, despite a Senior Vice President being informed, their failure to escalate the alert to their team allowed the oversight to persist (Kabanov & Madnick, 2020). By May 13, 2017, attackers exploited the unpatched system, moving laterally through Equifax's network and exfiltrating vast amounts of sensitive data (Diniyatullah & Rindu, 2024).

While there are few cybersecurity failures attributed to poor handoff communication in SOCs, we believe these well-known incidents represent only a small fraction of the many incidents and near misses resulting from handoff communication failures. Case studies from other high-risk industries may offer valuable insights into the potential consequences. The industries discussed below share a common challenge: when communication fails during shift changes or task handoffs, catastrophic outcomes can result. These examples illustrate the severe risks that could similarly apply to SOCs.

Healthcare

Healthcare provides numerous examples where failures in communication have led to delayed treatment, extended hospital stays, wrong-site surgeries, medication errors, and even mortality (The Joint Commission, 2017). In one incident, a failure in handoff communication nearly resulted in severe harm to a 29-year-old patient with brittle diabetes undergoing surgery. Critical details about his insulin sensitivity and recent hypoglycemia were lost when the transport assistant failed to relay them to the surgical team, and the electronic health record (EHR) lacked updated glucose data due to a documentation failure. As a result, the patient experienced severe hypoglycemia post-surgery, highlighting the dangers of communication breakdowns in high-risk clinical settings (Goldman & Catchpole, 2012).

When wrong-site surgeries occur, a majority of those instances can be attributed to failures in handoff communication (JCAHO Sentinel Event Statistics, 2003, as cited by Rogers et al., 2004.) The case of Willie King (Cook, 1998) involves a wrong-site surgery in which a patient's leg was amputated erroneously. Incident investigation revealed causes to be (1) the presence of two diseased legs, (2) paperwork that was updated but not shared among relevant team members, and (3) miscommunication among medical staff (Rogers et al., 2004). To mitigate incidents like wrong-site surgeries, the healthcare industry developed protocols and what were thought to be safeguards, such as marking surgical sites. However, ambiguity around whether an

"X" marked the limb to be operated on or the one to be left alone created confusion. This could parallel cybersecurity, where stakeholders often attempt to "patch" technical vulnerabilities involved in security incidents while neglecting to develop the non-technical skills of SOC personnel. Just as ambiguous markings can compound surgical risks rather than mitigate them, focusing exclusively on technical solutions rather than resolving underlying communication failures may end up introducing new vulnerabilities in SOCs.

Nuclear Power

The Sellafield nuclear site, burdened by a long history of nuclear operations, has experienced multiple incidents resulting in radioactive contamination of the surrounding environment, including its beaches (Vintro et al., n.d.). While pinpointing a singular "Sellafield Beach incident" is imprecise due to the site's complex history, the presence of radioactive particles and debris on adjacent beaches is a well-documented consequence of past operational failures and discharges (Brown & Etherington, 2011). Notably, a 1983 incident exemplifies the risks of flawed handoff procedures. In this case, a tank intended for ocean discharge was, in fact, filled with highly radioactive material, creating a substantial environmental hazard (Larder, 1996). Larder (1996) elucidates this handoff failure as a form of "telephone game," where logbook entries, written in shorthand across successive shifts, gradually distorted the tank's contents. Specifically, "ejections from HASW [High Activity Solid Waste]" were progressively misconstrued as "ex HASW washout." Consequently, material initially deemed unsuitable for ocean discharge was ultimately misidentified as low-level effluent, fit for release into the sea. This Sellafield incident underscores the acute dangers of deficient handoff processes within high-risk environments (Carvalho et al., 2012). Similarly, in SOCs, incomplete or ambiguous information transfer can lead to undetected threats, potentially amplifying the impact of cyberattacks.

Aviation

In aviation, the 1991 Continental Express Flight 2574 crash, caused by a maintenance shift handoff failure, illustrates how missing or overlooked information can lead to disastrous outcomes (National Transportation Safety Board [NTSB], 1991). A row of fasteners had been removed by personnel and not replaced during maintenance the night before the incident. The NTSB (1991) found that the error may have been detected if appropriate shift handoff procedures were completed. In cybersecurity, similarly, a failure to communicate key threat intelligence during a shift change could lead to an attack's success.

Oil and Gas

The BP Deepwater Horizon oil spill, a catastrophic environmental disaster, resulted from critical information gaps during shift handovers (Environmental Protection Agency [EPA], 2010). Specifically, vital reports regarding cement stability and OptiCem data were not effectively communicated to Transocean personnel on the

Deepwater Horizon rig or to the Houston office. This failure to transfer essential information contributed to the explosion, tragically claiming 11 lives and releasing an estimated 5 million barrels of oil into the Gulf of Mexico. In a SOC, incomplete handoffs can leave gaps in the response to ongoing incidents, potentially allowing attackers to exploit vulnerabilities.

SOC CONSIDERATIONS AROUND HANDOFFS

Effective handoffs are critical for maintaining continuous cybersecurity operations, yet they pose significant challenges, especially in the complex and global environment of modern SOCs. While the above sections explored what other fields with high consequences for failure are doing to make recommendations, we also want to introduce some special considerations that SOCs require around handoff strategies.

Reliance on Event Management Systems

Security information and event management (SIEM) systems collect and correlate log data from various sources (e.g., firewalls, servers, and intrusion detection systems). This facilitates monitoring, detecting, and alerting, allowing analysts to identify potential security incidents (Gürtürk & Aydın, 2024). Security Orchestration, Automation, and Response (SOAR) tools can enhance SIEM capabilities through the automation of tasks like incident triage and prioritization, containment (e.g., blocking IP addresses), and potentially gathering information to be included during handoff (Bridges et al., 2023).

SOCs face challenges despite event management systems, such as alert overload and integration issues, which can delay incident responses (Shankar & Madisetti, 2024). While event management systems (like SIEM and SOAR) are complementary to handoff processes, they are not a replacement. Event management systems provide data and automation capabilities, but they don't inherently dictate how that information is transferred between analysts. Standardized handoffs would provide that structure, to ensure critical information is consistently communicated. There is also sense-making to be done of raw data, even with automated processes. Handoff protocols ensure that logs, alerts, and incident details are given the proper context- the "why" behind the data. For example, SOAR can generate a report, but a handoff protocol is necessary to ensure that the nuances of complex situations and potential implications are understood by the receiver. Event management systems store data, while handoff protocols ensure that it's effectively transferred between analysts.

Challenges of Global SOC Handoffs

Much of the research on handoffs in the healthcare domain has focused on within-unit handoffs, such as shift changes. However, between-unit handoffs (those occurring when patients are transferred between institutions) present distinct challenges, including communication across different specialties, the absence of established relationships between sender and receiver, and a lack of shared situational awareness (Hilligoss & Cohen, 2013).

These challenges raise important questions about the complexities of handoffs in a global cybersecurity context. International SOC handoffs may face additional obstacles, particularly cultural differences, which have been identified as barriers to effective handoffs in other fields (Mascioli et al., 2009). The modern cybersecurity workforce consists of diverse teams working together to achieve common goals. Cultural differences can manifest during handoffs through differing communication styles, varying interpretations of sense of urgency, and time zone coordination challenges. All of these things can lead to omission or commission of information, such as loss of critical indicators of compromise, or incorrect incident timeline information being passed. Given these considerations, SOCs must carefully design their handoff protocols to address potential communication gaps, organizational differences, and cross-cultural factors that could impact the seamless transfer of critical security information.

Multiple Handoffs

In the context of critical security incidents, multiple handoffs may be necessary (Burg, 2020). This is similar to the military trauma system, where a single patient can undergo up to ten transfers of care (Bailey et al., 2013). Each handoff introduces the risk of information loss, which can lead to delays in treatment or errors (Arora & Johnson, 2006; Pezzolesi et al., 2010; Starmer et al., 2013). When adopting a "follow the sun" model, SOCs should be aware of the potential for data loss with each transfer and take steps to mitigate this risk. This risk can be compounded as incidents develop and escalate.

Balancing Confidentiality and Availability

In cybersecurity, the CIA triad stands for Confidentiality, Integrity, and Availability, which are core principles of security in information technology systems. Achieving a balance between these three is challenging, as prioritizing one can compromise the others. For example, information that is widely available may jeopardize its confidentiality and integrity, while highly confidential information can limit its availability (Aminzade, 2018).

The CIA triad is particularly relevant to handoff communications within SOCs. Unlike other domains, cybersecurity must place a high priority on privacy. SOCs must be careful in handling cybersecurity incident and event data to (1) comply with privacy policies related to their customers, and (2) prevent sensitive information from being compromised. SOCs also face internal security risks, such as insider threats, where malicious actors exploit legitimate access privileges (János & Dai, 2018). Wide information sharing strategies may unintentionally lead to data breaches or other cybersecurity incidents. The information shared during a SOC handoff is often confidential, especially during active threats. If compromised, details about an ongoing attack and mitigation efforts could be exploited by malicious actors.

Stress, Fatigue, and Burnout

SOC analysts face stress, fatigue, and burnout, which can impact their performance and retention (Dykstra & Paul, 2018; Nobles, 2022). Survey studies have shown that 44% of cybersecurity professionals experience severe work-related stress and burnout (Arora & Hastings, 2024). These issues arise from factors such as cognitive workload, unrealistic expectations, and organizational culture (Sundaramurthy et al., 2014; Arora & Hastings, 2024). Passalacqua and Segrin (2012) mention that all three of these factors significantly impact handoff quality in patient care settings. Solutions to this are complex. Reducing work hours can mitigate the risk of fatigue, but this also increases handoff frequency, meaning potentially more opportunities for information loss. In cybersecurity settings, recommendations include facilitating supportive work environments, implementing mindfulness programs, and addressing systemic challenges (Arora & Hastings, 2024).

RECOMMENDED HANDOFF PRACTICES

Here, we present some of the recommended handoff practices that have a proven track record.

Standardization

Effective handoffs ensure seamless transitions between shifts, preventing gaps in incident monitoring, detection, and response. Federal bodies, such as the Joint Commission, have advocated for the standardization of handoffs (Saufl, 2007). According to Riesenberg et al. (2009) standardized handoff process might look like:

- Checklist implementation
- Protocols
- Digital sign-out programs
- Mnemonics (see *Handoff Mnemonic for SOCs* below)

One standardized handoff process specifically put forth for SOCs is the SEAT-SWAP method (Crowley, 2021). More information on this mnemonic is discussed below. Importantly, meta-analytic evidence in healthcare supports the implementation of standardized handoffs, demonstrating positive effects across delivery of handoff information, patient outcomes, provider outcomes, and organizational outcomes (Keebler et al., 2016).

In regard to standardization, SOC teams should use the same (or highly similar) toolsets. Standardization of tooling can facilitate effective communication exchange during handoffs by eliminating confusion and ensuring that all SOC analysts are working from the same operational picture.

Explicit Transfer of Responsibility and Accountability

Shift-work is not as clearly defined as it once was with punch-in and punch-out time card systems. Active cybersecurity events or incidents may require that SOC analysts arrive ahead of time to receive a handoff or stay later to deliver one. Within healthcare settings, the clear and explicit transfer of responsibility is seen as an essential aspect of handoffs (Cooper, 2010; Patterson et al., 2004). Explicit accountability statements during handoffs reduce the likelihood of medical errors (Appelbaum et al., 2024), whereas unclear accountability contributes to a lack of ownership in patient care, leading to higher risk of error (Ong & Coiera, 2013). Clarification of responsibility ensures that all members of a team are aware of their roles, making for smoother transitions (Lazzari, 2024). SOC analysts may benefit from a clearly defined transition of responsibility, such as a written or verbal confirmation that the handoff was received.

Closed-Loop Communication

Closed-loop-communication ensures that the receiver is prepared, addresses questions the receiver might have, and confirms understanding through a readback (Fausett et al., 2024). This has been shown to alleviate the high incidence of miscommunication during handoffs in healthcare, and offers a standardized structure in which to conduct the verbal handoff report (Eva et al., 2023). A readback during a handoff looks like the receiver providing a summary of all the information they have captured from the receiver's handoff report. Readbacks ensure that all critical information is accurately conveyed, reducing the risk of miscommunication (Salik & Ashurst, 2019). The success of readbacks can be demonstrated by their wide use, in everything from military handoff operations (Knott et al., 2022) to placing an order at the drive-thru. Likewise, allocating time for questions creates a space to address any uncertainties and confirm details (Ahmed, 2022). As seen in space shuttle mission control operations, time for questions can be helpful not only to clarify information to the receiver, but also to uncover solutions to the issue at hand. Verifying questions, such as "Do you know that for sure?" can help detect fixation errors (Patterson et al., 2004).

HANDOFF MNEMONIC FOR SOCs

The guidance around handoff communication is largely domain agnostic. Things like closed-loop communication, explicit transfer of care, and time for questions are all relevant to ensuring a handoff goes smoothly, but do not dictate the information that should be included in the handoff itself. This information is idiosyncratic to the industry, organization, and domain. Mnemonics are commonly used to enhance memory. In the case of handoffs, mnemonics may increase memory of important steps and provide a structured process to follow. Sometimes, it is helpful for mnemonics to remind individuals engaged within the handoff process what information is helpful to include, although systematic literature review published in 2019 suggests that scientifically rigorous studies are needed to assess the effectiveness of handoff mnemonic (Riesenberg et al., 2019). There are at least 24 identified handoff mnemonics in healthcare, with SBAR (Situation, Background, Assessment, and Recommendation) being cited most frequently (Riesenberg et al., 2019).

Just one mnemonic, put forth by Chris Crowley in 2021, exists specifically for use within a SOC. This is the SEAT-SWAP (Staff, Explanation, Awareness, Transition,

Situation, Written, Accurate, and Persistent). Crowley expands on the SEAT portion of the acronym, with levels indicating what is "essential," what is "optimal" and what is "thorough." While SEAT offers some of the information that should be included, SWAP generally refers to the process surrounding the handoff (i.e., it should be about the situation, it should be written down, it should be accurate, and it should be done persistently). There seems to be a penchant for documentation in the security community. One Reddit post regarding handoffs in a SOC suggests that the only information that is okay to be provided verbally is when the last pot of coffee was brewed.

Staff

Staff refers to who should be performing the handoff. Both parties (sender and receiver) should be available to participate in the transition. Crowley argues that being physically present is important, and being mentally present is also important. "*Are you ready to receive a bunch of information upon arrival? If not, you're not the only one*" states Crowley, who argues that handoff discussions that occur within the first 30 minutes of shift-start is sub-optimal.

Explanation

Here, it is suggested to explain what is going on. For example, Crowley recommends:

- The active situation(s)
- The concerns
- Work already done
- Work to be done in the future

There are several tools that can help with this, and Crowley mentions slack plugins, checklists and forms, SOAR tools, and dedicated handoff tools. The emphasis here is on a genuine exchange of information from sender to receiver. While standardization remains valuable, Crowley notes that over-reliance on rigid checklists at the expense of genuine explanations of what matters and why can be counterproductive. Effective standardization can take many forms beyond checklists, including mnemonics, structured conversations, or consistent processes that still allow for meaningful information exchange. This nuance is often missed in handoff measurement in healthcare, where adherence to checklist items is frequently measured rather than the quality and effectiveness of the actual information transfer.

SOCs might benefit from adopting what healthcare calls "structured flexibility," using standardized frameworks that ensure critical information is consistently transferred while allowing space for contextual details and meaningful dialogue. For example, a SOC could implement a standardized SEAT-SWAP template that includes mandatory fields for tracking incidents and threats, but also dedicates time for narrative explanation and questions. Research is needed to determine the optimal balance of structure and conversation in cybersecurity handoffs, and how this balance might vary based on incident severity, team experience levels, and organizational context.

Awareness

When a problem occurs, information about that problem should be shared. There is also the importance of sharing information that could prevent a problem from occurring in the first place. According to Crowley (2021), awareness in a SOC means to guide future actions and decisions by bringing issues to light for consideration. These discussions should also incorporate briefings on operations and conditions pertaining to inside the system if these are known.

Transition

Crowley argues that shift changes are more effective if staff are already practicing continuity and consistent operational excellence. This means that the goal of a SOC is not to depend on the capabilities of any one individual but rather to consistently deliver service. This requires the development of standards, procedures, training, and information sharing before the handoff itself ever occurs. If this is not already the case, then implementing handoff procedures won't fix that. In fact, the shift change could be a cause of frustration due to inconsistencies.

The SWAP portion that follows is more of a values statement, see Table 5.3.

The strengths of the SEAT-SWAP method include an emphasis on presence, both physical and mental. This is important as both parties should be engaged during a handoff (distractions can lead to information loss). The critiques on over-reliance on checklists are valid; structured tools can improve standardization but should not replace conversational exchanges that establish context and priority. The focus on system wide consistency rather than individual competency aligns with the best practices for handoffs that we see in healthcare.

TABLE 5.3
SWAP Portion of SEAT-SWAP

Situation	Describes the situation at hand.
Written	The handoff should be in a written form, and Crowley suggests that a recorded briefing that could be reviewed later is also helpful. Some prefer to read, others prefer to listen. This can change depending on the topic and your team's attention bandwidth.
Accurate	The handoff should be in a written form, and Crowley suggests that a recorded briefing that could be reviewed later is also helpful. Some prefer to read, others prefer to listen. This can change depending on the topic and your team's attention bandwidth.
Persistently	Crowley says that shift changes should be done frequently. In his words, "this is not something that can be done *sometimes* or *as needed*." The shift change always supports, even if there is a "nothing to report" statement.

Source: Crowley (2021).

VERBAL HANDOFF EXAMPLE

Here is an example of verbal handoff that incorporates some of the recommended handoff practices above, and uses the SEAT-SWAP method introduced by Crowley (2021).

Sender: Are you ready to receive the handoff?

Receiver: Yes, I am ready.

Sender: We detected and blocked an attempted credential stuffing attack targeting our customer portal. The source IPs were primarily from a known botnet range, and we've confirmed failed login attempts across 1,200 different accounts over a 15-minute window. MFA stopped the majority, but at least three accounts appear to have successfully logged in right after multiple failed attempts, which is unusual. We've locked those accounts and sent alerts to the users. No lateral movement detected, and there's no sign yet of data exfiltration. The firewall logs are clean apart from this activity, and we've temporarily geo-blocked the region where most of the IPs originated. No immediate customer complaints, but we should keep an eye on support tickets. The SIEM flagged similar activity from a different set of IPs last week, but this time the scale was bigger.

Receiver: Alright, let me make sure I've got this right. You detected and blocked a credential stuffing attack targeting the customer portal. The attack used IPs from a known botnet, and there were failed login attempts on around 1,200 accounts within 15 minutes. MFA stopped most of them, but you saw three accounts log in successfully right after multiple failures, which is suspicious. You've locked those accounts and sent alerts to the users. No lateral movement so far, no data exfiltration detected, and the SIEM flagged similar activity from a different set of IPs last week.

Sender: That's mostly right, but don't forget we also geo-blocked the region where most of the attack traffic came from. We should keep an eye on whether that causes any false positives for legitimate users.

Receiver: Got it—you geo-blocked the region where most of the attack traffic came from. I'll watch for any false positives in case legitimate users get blocked.

Sender: Yep, that's it on my end. Any questions?

Receiver: Do we know if any of the compromised accounts had admin privileges?

Sender: I wasn't able to verify before our shift-change, but we haven't received any customer complaints yet.

Receiver: Alright, I'll make checking for admin privileges my first priority. This situation is mine.

Sender: The situation is yours.

Our example does not include a written component, but we encourage one to imagine what this might look like as this is an incredibly important aspect of the handoff. The written component would likely be documented in a SIEM log or a SOAR tool. These systems are designed to capture and organize details of security incidents, including actions taken and outstanding tasks. Alternatively, some SOCs anecdotally favor Slack messages or Email for written communications. This verbal handoff scenario aligns nicely with the SEAT portion of Crowley's method:

(S) Staff are prioritized as the handoff is framed in a way where both the sender and receiver are present and mentally prepared to exchange information. This is facilitated by closed-loop communication, wherein the beginning of the handoff asks if both parties are ready to participate in the handoff. The shift transition isn't rushed, and the receiver's acknowledgment that they will prioritize checking for admin privileges demonstrates engagement.

(E) Explanation is clear on the sender's behalf, explaining the situation, outlining key details about the attack (credential stuffing, botnet origin, MFA stopping most of the attacks, geo-blocking actions) and what has been done (locking the accounts, sending alerts, SIEM flagging activity). The mention of "no customer complaints yet" and the detail about "not being able to verify admin privileges" highlights both current actions and future steps.

(A) Awareness is clear from the conversation. For example, when the sender emphasizes the need to check for admin privileges, they are sharing information that could prevent a problem from occurring (compromised accounts with elevated access). The acknowledgment that geo-blocking could cause false positives is also proactive, providing the receiver with insight into potential future issues. This is in line with Endsley's (2000) levels of situation awareness: level 1 (perception) is demonstrated through the detailed description of the attack and work done, level 2 (comprehension) is shown when the receiver considers the situation and prioritizes their next action (checking admin access), and level 3 (projection) is seen in the proactive identification of possible future issues like false positives.

(T) Transition is continuous and conducted in a way that minimizes frustration. The receiver knows exactly what steps are needed next, such as verifying admin access and watching for false positives from geo-blocking. This smooth transition reflects the importance of having standards and procedures in place for effective shift changes. The receiver is not left guessing or scrambling for information but is guided toward the next action, with a clear understanding of the situation.

The handoff in this scenario goes beyond just a checklist and emphasizes genuine information exchange, promoting shared awareness of the situation, and ensuring the continuity of operations in the SOC. It demonstrates a well-structured handoff that aligns with Crowley's SEAT method principles, fostering effective collaboration and a seamless transition between shifts.

Importantly, it also follows the recommended practices for handoffs that have been gleaned from other domains:

Standardization

By following a structured approach (the SEAT-SWAP method), each part of the exchange is clearly demarcated, ensuring both parties understand the immediate situation and the next steps. The handoff example demonstrates the importance of standardization in SOC operations by ensuring both the sender and receiver are working from the same operational picture.

Explicit Transfer of Responsibility and Accountability

In the handoff, the sender makes it clear that they weren't able to verify whether the compromised accounts had admin privileges, and they specifically pass that responsibility on to the receiver. The receiver acknowledges the task ("I'll make checking for admin privileges my first priority"), which confirms that responsibility for that particular aspect of the investigation has been transferred. Finally, an explicit transfer occurs at the end of the handoff, with statements of "the situation is mine" and "the situation is yours."

Closed-Loop Communication

The handoff example above contains both a readback and time for questions. The receiver demonstrates closed-loop communication by repeating back the key details of the handoff and confirming their understanding. When they say, "I'll make checking for admin privileges my first priority," they are not only acknowledging what needs to be done but are also clarifying that they understand the importance of that task. The handoff leaves room for the receiver to ask clarifying questions, which revealed a gap in the senders' initial communication regarding if any of the compromised accounts had admin privileges. This is a high-priority item that was almost missed, but was able to be transferred thanks to closed-loop communication.

RESEARCH AGENDA

Holistic Understanding of Current SOC Handoff Operations

Without an understanding of current SOC handoff operations, it is difficult to adequately suggest interventions. While the guidance above is rooted in best practices and available knowledge, it lacks input from subject-matter experts. Future research should seek to overcome interdisciplinary barriers and explore (1) how analysts conduct handoffs, (2) what information is transferred (both verbally and through written documentation), (3) the most critical challenges surrounding SOC handoffs. Comparative analyses conducted across multiple SOCs could help shed light on successful attributes of SOC handoffs before experimental validation would be able to occur.

The Impact of Stress, Fatigue, and Burnout

Stress, fatigue, and burnout have documented impacts on handoffs and decision-making in high-pressure environments, but there has been little research on how these factors impact handoff operations in a SOC. Future research could consider how these factors influence handoff effectiveness, and the extent to which analysts are capable of making situation assessments. Research from industrial organizational and human factors psychology can help create interventions.

Applicability of Recommended Handoff Practices

Although this chapter argues that recommended practices for handoffs will likely be applicable to SOCs, this remains an important question to address through rigorous research. As noted in this chapter, there are unique considerations important to SOC operations that will likely necessitate the development and implementation of new SOC-specific handoff strategies. The development and implementation of these strategies would benefit from qualitative and exploratory research methods to identify context-specific needs and processes, while subsequent validation and effectiveness testing could employ experimental designs to evaluate their impact on SOC performance and outcomes.

The Impact of Artificial Intelligence and Automation

Artificial intelligence and automation are rapidly becoming inseparable from SOC operations. Future work could examine the role of these tools in the handoff process. Artificial intelligence may certainly boost efficiency and lower cognitive demands on SOC analysts. Similarly, over-reliance on automation may lead to a reduction in situation awareness, or introduce new risks. Research could investigate handoff performance in SOCs that are largely integrated with AI tools compared to SOCs that are not and rely primarily on manual flagging and documentation.

Training and Measurement of SOC Handoff Operations

Training interventions are also an unexplored area. While the recommended practices here have been validated in other domains, their effectiveness in cybersecurity settings remains unknown. Other domains have success with handoff training simulations, training sessions, and scenario-based exercises. However, in order to effectively evaluate training, team performance and handoff outcome metrics are needed. While some metrics for SOC performance exist, these may not reflect the human aspect of the larger sociotechnical system. One cannot improve without measurement.

The Influence of Organizational Culture

Organizational culture is also a very important part of team processes, and therefore handoff operations. Leadership, team dynamics, and aspects like psychological safety can all influence the transfer of information during SOC handoff operations. SOC case studies could be a valuable place to start establishing the importance of organizational culture.

CONCLUSION

This chapter has provided a practical, evidence-based examination of SOC handoffs by drawing on established knowledge from other domains, particularly healthcare. While limited research exists specifically on cybersecurity handoffs, the chapter offers actionable insights and best practices that can be immediately applied to improve SOC operations. Beyond these practical contributions, the chapter concludes with a call to action for applied researchers: cross-domain collaboration is essential to further refine and validate handoff strategies in cybersecurity contexts. Future research should explore current handoff operations, the effects of stress, fatigue, and burnout, the applicability of recommended handoff practices to SOCs, the impacts of artificial intelligence and automation, and the influence of organizational culture. This interdisciplinary approach will help advance both the theory and practice of effective handoffs in the rapidly evolving cybersecurity landscape.

REFERENCES

Ahmed, I. (2022). Clinical Hand Over: A Neglected Domain in Our Health Care. ISRA Medical Journal, 14(2), 48–49. https://doi.org/10.55282/imj.ed60

Alashwali, E., Mysuru Chandrashekar, R., Lanyon, M., & Faith Cranor, L. (2024, September). Detection and Impact of Debit/Credit Card Fraud: Victims' Experiences. In *Proceedings of the 2024 European Symposium on Usable Security* (pp. 235–260).

Aminzade, M. (2018). Confidentiality, Integrity and Availability–Finding a Balanced IT Framework. Network Security, 2018(5), 9–11.

Appelbaum, R. D., Puzio, T. J., Bauman, Z., Asfaw, S., Spencer, A., Dumas, R. P., … Freeman, J. (2024). Handoffs and Transitions of Care: A Systematic Review, Meta-analysis, and Practice Management Guideline from the Eastern Association for the Surgery of Trauma. Journal of Trauma and Acute Care Surgery, 97(2), 305–314.

Arora, S., & Hastings, J. D. (2024). A survey-based quantitative analysis of stress factors and their impacts among cybersecurity professionals. arXiv preprint arXiv:2409. 12047.

Arora, V., & Johnson, J. (2006). A model for Building a Standardized Hand-off Protocol. The Joint Commission Journal on Quality and Patient Safety, 32(11), 646–655.

Arora, A., N., Kumar, A., & Telang, R. (2006). Does Information Security Attack Frequency Increase with Vulnerability Disclosure? An Empirical Analysis. Information Systems Frontiers, 8(5), 350–362.

Ashforth, B. E., & Mael, F. (1989). Social Identity Theory and the Organization. Academy of Management Review, 14(1), 20. https://doi.org/10.2307/258189

Bailey, C. J. A., Morrison, M. J. J., & Rasmussen, C. T. E. (2013). Military Trauma System in Afghanistan: Lessons for Civil Systems? Current Opinion in Critical Care, 19(6), 569–577.

Benishek, L. E., & Lazzara, E. H. (2019). Teams in a New Era: Some Considerations and Implications. Frontiers in Psychology, 10. https://doi.org/10.3389/fpsyg.2019.01006

Berezin, J. (2025, January 20). How to build a security operations center, and why you need one. Cyop Security. Retrieved March 2, 2025, from https://cyopsecurity.com/insights/how-to-build-a-security-operations-center-and-why-you-need-one/

Beyond Machines. (2022, July 10). I quit!—Cybersecurity burnout [Blog]. Medium. https://medium.com/@beyondmachines/i-quit-cybersecurity-burnout-c42c04cb0d53

Boyer, S. (2019, October 11). We're all at risk when 65% of stressed-out cybersecurity and IT workers are thinking about quitting, tech exec warns. CNBC. https://www.cnbc.com/2019/10/11/65percent-of-stressed-out-cybersecurity-it-workers-think-about-quitting.html

Bridges, R. A., Rice, A. E., Oesch, S., Nichols, J. A., Watson, C., Spakes, K., … Erwin, S. (2023). Testing SOAR Tools in Use. Computers & Security, 129, 103201.

Brilingaitė, A., Bukauskas, L., Juozapavičius, A., & Kutka, E. (2022). Overcoming Information-sharing Challenges in Cyber Defence Exercises. Journal of Cybersecurity, 8(1), tyac001.

Brown, J., & Etherington, G. (2011). Health risks from radioactive objects on beaches in the vicinity of the Sellafield site (HPA-CRCE-018). Health Protection Agency.

Burg, S. (2020, July 28). Follow the sun model - Improve remote cybersecurity training. Cyberbit. Retrieved February 6, 2022, from https://www.cyberbit.com/blog/cybersecurity-training/follow-the-sun-model-improve-remote-cybersecurity-training/

Burke, C., Fiore, S., & Salas, E. (2003). The role of shared cognition in enabling shared leadership and team adaptability. In C. L. Pearce, J. A. Conger (Eds.), The role of shared cognition in enabling shared leadership and team adaptability (pp. 103–122). SAGE Publications, Inc. https://doi.org/10.4135/9781452229539.n5

Bykowski, A. (2023, March 29). SOC team roles & responsibilities. Swimlane. Retrieved March 2, 2025, from https://swimlane.com/blog/soc-team-roles-responsibilities/

Carvalho, P. V. R. D., Benchekroun, T., & Gomes, J. O. (2012). Analysis of Information Exchange Activities to Actualize and Validate Situation Awareness during Shift Changeovers in Nuclear Power Plants. Human Factors and Ergonomics in Manufacturing & Service Industries, 22(2), 130–144. https://doi.org/10.1002/hfm.20201

Cassetto. (2022). Security operations center roles and responsibilities. Exabeam. Retrieved March 2, 2025, from https://www.exabeam.com/blog/security-operations-center/security-operations-center-roles-and-responsibilities/

Chamkar, S. A., Maleh, Y., & Gherabi, N. (2024). Security Operations Centers: Use Case Best Practices, Coverage, and Gap Analysis Based on MITRE Adversarial Tactics, Techniques, and Common Knowledge. Journal of Cybersecurity and Privacy, 4(4), 777–793. https://doi.org/10.3390/jcp4040036

Champion, M., Rajivan, P., Cooke, N. J., & Jariwala, S. (2012). Team-based cyber defense analysis. *2012 IEEE International Multi-Disciplinary Conference on Cognitive Methods in Situation Awareness and Decision Support* (pp. 218–221).

Cichonski, P., Millar, T., Grance, T., & Scarfone, K. (2012). Computer security incident handling guide (NIST Special Publication 800-61, Rev. 2). National Institute of Standards and Technology. https://doi.org/10.6028/NIST.SP.800-61r2

Condello, T. (2020, June 17). How to handle shift handovers in the Security Operations Center. Siemplify. Retrieved February 6, 2022, from https://www.siemplify.co/blog/how-to-handle-shift-handovers-in-the-security-operations-center/

Cook, R. I. (1998). How Complex Systems Fail. In Cognitive technologies laboratory (pp. 64–118.), Chicago: University of Chicago.

Cooper, J. B. (2010). Using Simulation to Teach and Study Healthcare Handoffs. Simulation in Healthcare, 5(4), 191–192.

Crowley, C. (2021, November 17). Shifts happen: How to rock the SOC handoff process with the seat swap method. Security Boulevard. https://securityboulevard.com/2021/11/shifts-happen-how-to-rock-the-soc-handoff-process-with-the-seat-swap-method/

Crowley, C. (2024). SANS 2024 SOC Survey: Facing Top Challenges in Security Operations. SANS Institute. https://www.sans.org/white-papers/sans-2024-soc-survey-facing-top-challenges-security-operations

Diniyatullah, L., & Rindu, K. B. (2024). Crisis Management and Recovery Strategies After a Data Leak: Equifax Case Study. JOISTECH: Journal of Information System and Technology, 1(2), 76–81.

Driskell, T., Funke, G., Tolston, M., Capiola, A., & Driskell, J. E. (2023). Fluid and Virtual Teams. Small Group Research, 55(5), 680–705. https://doi.org/10.1177/10464964231216367

Dykstra, J., & Paul, C. L. (2018). Cyber Operations Stress Survey (COSS): Studying fatigue, frustration, and cognitive workload in cybersecurity operations. In *11th USENIX Workshop on Cyber Security Experimentation and Test (CSET 18).*

Endsley, M. R. (2000). *Theoretical underpinnings of situation awareness: A critical review.* In M. R. Endsley & D. J. Garland (Eds.), *Situation awareness analysis and measurement* (pp. 3–33). CRC Press.

Environmental Protection Agency (EPA). (2010). BP Deepwater Horizon oil spill: technical assessment and response (EPA Report No. 430-R-10-005). Washington, DC: U.S. Environmental Protection Agency.

Equifax. (2020). Data Breach Settlement. https://www.equifaxbreachsettlement.com/

Etschmaier, M. M. (2019). Critical Issues of Cybersecurity: Solutions beyond the Technical. International Journal of Computer Applications, 26, 140–153.

Eva, C., Slim, N., Ghatge, A., Bruno, P., Blackaby, J., Cutting, J., & Fowler, G. E. (2023). 861 Getting the Surgical Handover Right: A Closed-Loop Audit. British Journal of Surgery, 110(1). https://doi.org/10.1093/bjs/znad258.150

Fausett, C. M., & Keebler, J. R. (2022, September). Shift Happens: Human Factors Considerations for Handoffs in Cyber-security Operations Centers. In *Proceedings of the Human Factors and Ergonomics Society Annual Meeting* (Vol. 66, No. 1, pp. 110–114). Los Angeles, CA: SAGE Publications.

Fausett, C. M., Korentsides, J. M., Knott, C. C., Chik, K., Grome, A., Sciarini, L. W., & Keebler, J. R. (2024). Redesign of Military Patient Handoff Training for Emergency Medical Services. *Ergonomics in Design: The Quarterly of Human Factors Applications*, *33*(2), 73–80. https://doi.org/10.1177/10648046241237086 (Original work published 2025).

Furnell, S., Heyburn, H., Whitehead, A., & Shah, J. N. (2020). Understanding the Full Cost of Cyber Security Breaches. Computer Fraud & Security, *2020*(12), 6–12.

Gao, R., Peng, C., Wang, X., & Wan, Z. (2024). Psychological Safety Climate and Self-managing Team Innovation Process: The Role of Constructive Controversy and Authority Differentiation. Group Dynamics Theory Research and Practice. https://doi.org/10.1037/gdn0000214

Ghafur, S., Kristensen, S., Honeyford, K., Martin, G., Darzi, A., & Aylin, P. (2019). A Retrospective Impact Analysis of the WannaCry Cyberattack on the NHS. NPJ Digital Medicine, 2(1), 98.

Goldman, A., & Catchpole, K. (2012, September 1). Empty handoff. Agency for Healthcare Research and Quality. https://psnet.ahrq.gov/web-mm/empty-handoff

Gürtürk, U., & Aydın, Z. G. (2024). An Incident Management System Design to Protect Critical Infrastructures from Cyber Attacks. Journal of Mathematical Sciences and Modelling, 7(2), 60–74.

Hackman, J. R., & Katz, N. (2010). Group behavior and performance. In S. T. Fiske, D. T. Gilbert, & G. Lindzey (Eds.), Handbook of social psychology (pp. 1208–1251). John Wiley & Sons, Inc. Hoboken, NJ, USA. https://doi.org/10.1002/9780470561119.socpsy002032

Hámornık, B. P., & Krasznay, C. (2018). A team-level perspective of human factors in cyber Security: Security operations centers. In D. Nicholson (Ed.), Advances in human factors in cybersecurity (Vol. 593, pp. 224–236). Cham, Switzerland. Springer International Publishing. https://doi.org/10.1007/978-3-319-60585-2_21

Hilligoss, B., & Cohen, M. D. (2013). The Unappreciated Challenges of Between-unit Handoffs: Negotiating AND Coordinating Across Boundaries. Annals of Emergency Medicine, 61(2), 155–160.

Hollenbeck, J. R., Beersma, B., & Schouten, M. E. (2012). Beyond Team Types and Taxonomies: A Dimensional Scaling Conceptualization for Team Description. The Academy of Management Review, 37(1), pp. 82–106.

Institute of Medicine (US) Committee on Quality of Health Care in America, Kohn, L. T., Corrigan, J. M., & Donaldson, M. S. (Eds.). (2000). To err is human: Building a safer health system. US: National Academies Press.

Ioannou, M., Stavrou, E., & Bada, M. (2019). Cybersecurity culture in computer security incident response teams: Investigating difficulties in communication and coordination. *2019 International Conference on Cyber Security and Protection of Digital Services (Cyber Security)*, 1–4.

János, F. D., & Dai, N. H. P. (2018, May). Security concerns towards security operations centers. In *2018 IEEE 12th International Symposium on Applied Computational Intelligence and Informatics (SACI)* (pp. 000273–000278). IEEE.

Kabanov, I., & Madnick, S. E. (2020). A systematic study of the control failures in the Equifax cybersecurity incident. MIT Sloan Research Paper No. 2020-19. http://dx.doi.org/10.2139/ssrn.3957272

Kane, A. A., Argote, L., & Levine, J. M. (2004). Knowledge Transfer between Groups via Personnel Rotation: Effects of Social Identity and Knowledge Quality. Organizational Behavior and Human Decision Processes, 96(1), 56–71. https://doi.org/10.1016/j.obhdp.2004.09.002

Kaplan, D. (2023, June 29). How to nail smooth shift handoffs in SecOps. Google Cloud Blog. https://cloud.google.com/blog/products/identity-security/how-to-nail-smooth-shift-handoffs-in-secops

Keebler, J. R., Lazzara, E. H., Patzer, B. S., Palmer, E. M., Plummer, J. P., Smith, D. C., … Riss, R. (2016). Meta-analyses of the Effects of Standardized Handoff Protocols on Patient, Provider, and Organizational Outcomes. Human Factors, 58(8), 1187–1205.

Knerler, K., Parker, I., & Zimmerman, C. (2022). 11 Strategies of a World-Class Cybersecurity Operations Center.

Knott, C. K., Keebler, J. R., Fausett, C. M., Chik, K., Rudinsky, S., Conley, S., Jeyarajah, T., Lopreiato, J., Sciarini, L. W., & Wilson, R. (2022). Patient handoffs in combat environments: Training effectiveness and evaluation with TCCC Tier 1 and Tier 2 Equivalent Trainees. Military Health System Research Symposium, Orlando, FL.

Koopmann, J., Lanaj, K., Wang, M., Zhou, L., & Shi, J. (2016). Nonlinear Effects of Team Tenure on Team Psychological Safety Climate and Climate Strength: Implications for Average Team Member Performance. Journal of Applied Psychology, 101(7), 940–957. https://doi.org/10.1037/apl0000097

Kozlowski, S. W., & Ilgen, D. R. (2006). Enhancing the Effectiveness of Work Groups and Teams. Psychological Science in the Public Interest, 7(3), 77–124. https://doi.org/10.1111/j.1529-1006.2006.00030.x

Kwon, H. C., Chung, B. H., Moon, D. S., & Kim, I. K. (2021). Security Technology Trends to Prevent Medical Device Hacking and Ransomware. Electronics and Telecommunications Trends, 36(5), 21–31.

Larder, R. (1996). OTO 976 003 - Effective shift handover - A literature review. Health and Safety Executive.

Larson, J. R. Jr, & Schaumann, L. J. (1993). Group Goals, Group Coordination, and Group Member Motivation. Human Performance, 6(1), 49–69. https://doi.org/10.1207/s15327043hup0601_3

Lazzari, C. (2024). Implementing the Verbal and Electronic Handover in General and Psychiatric Nursing using the Introduction, Situation, Background, Assessment, and Recommendation Framework: A Systematic Review. Iranian Journal of Nursing and Midwifery Research, 29(1), 23–32.

Lin, C., He, H., Baruch, Y., & Ashforth, B. E. (2016). The Effect of Team Affective Tone on Team Performance: The Roles of Team Identification and Team Cooperation. Human Resource Management, 56(6), 931–952. https://doi.org/10.1002/hrm.21810

Litchfield, R. C., Karakitapoğlu-Aygün, Z., Gumusluoglu, L., Carter, M., & Hirst, G. (2017). When Team Identity Helps Innovation and When It Hurts: Team Identity and Its Relationship to Team and Cross-Team Innovative Behavior. Journal of Product Innovation Management, 35(3), 350–366. https://doi.org/10.1111/jpim.12410

Marks, M. A., Mathieu, J. E., & Zaccaro, S. J. (2001). A Temporally Based Framework and Taxonomy of Team Processes. Academy of Management Review, 26(3), 356–376. https://doi.org/10.5465/amr.2001.4845785

Mascioli, S., Laskowski-Jones, L., Urban, S., & Moran, S. (2009). Improving Handoff Communication. Nursing 2024, 39(2), 52–55.

Merriam-Webster (n.d.). Team. In *Merriam-Webster.com dictionary*. Retrieved January 23, 2025 from https://www.merriam-webster.com/dictionary/team

Miller, J. (2024, October 23). The importance of security automation in financial services. BitLyft. Retrieved March 2, 2025 from https://www.bitlyft.com/resources/the-importance-of-security-automation-in-financial-services

Miloslavskaya, N. (2016, August). Security operations centers for information security incident management. In *2016 IEEE 4th International Conference on Future Internet of Things and Cloud (FiCloud)* (pp. 131–136). IEEE.

National Institute of Standards and Technology (NIST). (2012). Computer security incident handling guide (NIST Special Publication 800-61 Rev. 2). https://csrc.nist.gov/pubs/sp/800/61/r2/final

National Transportation Safety Board (NTSB). (1991). Aircraft accident report: Continental Express Flight 2574 (Report No. NTSB/AAR-91/07). U.S. Government Printing Office.

Nilsen, M., Rasmussen, M., & Røyrvik, J. (2018, June). Handing Over the Safety Baton in High-Risk Systems. In International Conference on Applied Human Factors and Ergonomics (pp. 501–509). Cham: Springer International Publishing.

Nobles, C. (2022). Stress, burnout, and security fatigue in cybersecurity: A human factors problem. HOLISTICA – Journal of Business and Public Administration, 13(1), 49–72. https://doi.org/10.2478/hjbpa-2022-0003

Nominet. (2019). Life inside the perimeter: Understanding the modern CISO. Nominet. https://media.nominet.uk/wp-content/uploads/2019/02/12130924/Nominet-Cyber_CISO-report_FINAL-130219.pdf

Ong, M. S., & Coiera, E. (2013). Handoff and care transitions. In A. Agrawal (Ed.), Patient safety: A case-based comprehensive guide (pp. 35–51). New York, NY: Springer New York.

Passalacqua, S. A., & Segrin, C. (2012). The Effect of Resident Physician Stress, Burnout, and Empathy on Patient-Centered Communication during the Long-call Shift. Health Communication, 27(5), 449–456. https://doi.org/10.1080/10410236.2011.606527

Patterson, E. S., Roth, E. M., Woods, D. D., Chow, R., & Gomes, J. O. (2004). Handoff Strategies in Settings with High Consequences for Failure: Lessons for Health Care Operations. International Journal for Quality in Health Care, 16(2), 125–132.

Patterson, E. S., & Wears, R. L. (2009). Beyond "Communication Failure." Annals of Emergency Medicine, 53(6), 711.

Patterson, E. S., & Wears, R. L. (2010). Patient Handoffs: Standardized and Reliable Measurement Tools Remain Elusive. The Joint Commission Journal on Quality and Patient Safety, 36(2), 52–61.

Pezzolesi, C., Schifano, F., Pickles, J., Randell, W., Hussain, Z., Muir, H., & Dhillon, S. (2010). Clinical Handover Incident Reporting in one UK General Hospital. International Journal for Quality in Health Care, 22(5), 396–401.

Plutarch. (1994). Theseus (J. Dryden, Trans.). The Internet Classics Archive https://classics.mit.edu/Plutarch/theseus.html (Original work published ca. 75 A.C.E.)

Reagans, R., Argote, L., & Brooks, D. (2005). Individual Experience and Experience Working Together: Predicting Learning Rates from Knowing Who Knows What and Knowing How to Work Together. Management Science, 51(6), 869–881. https://doi.org/10.1287/mnsc.1050.0366

Reeves, A., Pattinson, M., & Butavicius, M. (2023). Is Your CISO Burnt Out Yet? Examining Demographic Differences in Workplace Burnout amongst Cyber Security Professionals. Human Aspects of Information Security and Assurance, 2023, 225–236. https://doi.org/10.1007/978-3-031-38530-8_18

Riesenberg, L. A., Leitzsch, J., & Little, B. W. (2009). Systematic Review of Handoff Mnemonics Literature. American Journal of Medical Quality, 24(3), 196–204.

Riesenberg, L. A., Leitzsch, J., & Little, B. W. (2019). Republished: Systematic Review of Handoff Mnemonics Literature. American Journal of Medical Quality, 34(5), 446–454. https://doi.org/10.1177/1062860619873200

Rogers, M. L., Cook, R. I., Bower, R., Molloy, M., & Render, M. L. (2004). Barriers to Implementing Wrong Site Surgery Guidelines: A Cognitive Work Analysis. IEEE Transactions on Systems, Man, and Cybernetics - Part A: Systems and Humans, 34(6), 757–763. https://doi.org/10.1109/TSMCA.2004.836805

Saavedra, R., Earley, P. C., & Van Dyne, L. (1993). Complex Interdependence in Task-performing Groups. Journal of Applied Psychology, 78(1), 61.

Salas, E., Dickinson, T. L., Converse, S. A., & Tannenbaum, S. I. (1992). Toward an understanding of team performance and training. In R. W. Swezey & E. Salas (Eds.), Teams: Their training and performance (pp. 3–29). Norwood, NJ: Ablex.

Salas, E., & Fiore, S. M. (Eds.). (2004). Team cognition: Understanding the factors that drive process and performance. American Psychological Association. https://doi.org/10.1037/10690-000

Salas, E., Stagl, K. C., Burke, C. S., & Goodwin, G. F. (2007). Fostering team effectiveness in organizations: Toward an integrative theoretical framework. In B. Shuart, W. Spaulding, & J. Poland (Eds.), Modeling complex systems (Vol. 52, pp. 185–243). University of Nebraska Press.

Salik, I., & Ashurst, J. V. (2025). *Closed Loop Communication Training in Medical Simulation*. In *StatPearls [Internet]*. StatPearls Publishing. https://europepmc.org/article/NBK/nbk549899.

Saufl, N. M. (2007). 2007 National Patient Safety Goals. Journal of PeriAnesthesia Nursing, 22(2), 125–127. https://doi.org/10.1016/j.jopan.2007.01.006

The Joint Commission. (2017, September 12). Sentinel event alert 58: Inadequate hand-off communication. https://www.jointcommission.org/resources/sentinel-event/sentinel-event-alert-newsletters/sentinel-event-alert-58-inadequate-hand-off-communication/

Shackelford, S. J. (2014). Managing cyber attacks in international law, business, and relations: In search of cyber peace. New York, USA: Cambridge University Press,.

Shankar, A., & Madisetti, V. (2024). A Framework for Cybersecurity Alert Distribution and Response Network (ADRIAN). Journal of Software Engineering and Applications, 17(5), 396–420.

Simonson, R. J., Keebler, J. R., Lessmiller, M., Richards, T., & Lee, J. C. (2020, December). Cybersecurity Teamwork: A Review of Current Practices and Suggested Improvements. In *Proceedings of the Human Factors and Ergonomics Society Annual Meeting* (Vol. 64, No. 1, pp. 451–455). Los Angeles, CA: SAGE Publications.

Stagl, K. C., Salas, E., Rosen, M. A., Priest, H. A., Burke, C. S., Goodwin, G. F., & Johnston, J. H. (2007). Distributed Team Performance: A Multi-Level Review of Distribution, Demography, and Decision Making. Research in Multi-level Issues, 11–58. https://doi.org/10.1016/s1475-9144(07)06001-8

Starmer, A. J., Sectish, T. C., Simon, D. W., Keohane, C., McSweeney, M. E., Chung, E. Y., … Landrigan, C. P. (2013). Rates of Medical Errors and Preventable Adverse Events among Hospitalized Children Following Implementation of a Resident Handoff Bundle. JAMA, 310(21), 2262–2270.

Sumner, B. D., Grimsley, E. A., Cochrane, N. H., Keane, R. R., Sandler, A. B., Mullan, P. C., & O'Connell, K. J. (2019). Videographic Assessment of the Quality of EMS to ED Handoff Communication during Pediatric Resuscitations. Prehospital Emergency Care, 23(1), 15–21.

Sundaramurthy, S. C., McHugh, J., Ou, X. S., Rajagopalan, S. R., & Wesch, M. (2014). An anthropological approach to studying CSIRTs. IEEE Security & Privacy, 12(5), 52–60.

Thompson, J. D. (1967). Organizations in action. New York: McGraw Hill.

Van de Ven, A. H., Delbecq, A. L., & Koenig, R. (1976). Determinants of Coordination Modes within Organizations. American Sociological Review, 41, 322–338.

Vielberth, M., Böhm, F., Fichtinger, I., & Pernul, G. (2020). Security Operations Center: A Systematic Study and Open Challenges. IEEE Access, 8, 227756–227779.

Vintró, L. L., Smith, K. J., Lucey, J. A., & Mitchell, P. I. (2000, December). The environmental impact of the Sellafield discharges. In SCOPE-RADSITE Workshop Proceedings, Brussels (pp. 4–6).

Wang, J., Liu, J., Sun, Q., Yan, T., An, D., & Qi, F. (2023). A Distributed Framework for Security Operation Center in the Application of Institute of High Energy Physics. *Proceedings of International Symposium on Grids & Clouds (ISGC) 2023 in Conjunction with HEPiX Spring 2023 Workshop — PoS*(ISGC&HEPiX2023), 026. https://doi.org/10.22323/1.434.0026

Weiss, M., & Hoegl, M. (2015). Effects of Relative Team Size on Teams with Innovative Tasks. Organizational Psychology Review, 6(4), 324–351. https://doi.org/10.1177/2041386615620837

Weissman, D. L., & Jayasumana, A. P. (2020). Integrating IoT Monitoring for Security Operation Center. 2020 Global Internet of Things Summit (GIoTS), 1–6.

Widianto, S., Sahib, H. M. A., & Rahman, M. F. W. (2024). Task Interdependence, Team Identity and Team Performance: A Bottom-Up Multilevel Model. SAGE Open, 14(1). https://doi.org/10.1177/21582440241237874

Zaccaro, S. J., Dalal, R. S., Tetrick, L. E., & Steinke, J. A. (2016). The psychosocial dynamics of cybersecurity. In Zaccaro S. J., Dalal, R. S., Tetrick, L. E., Steinke, J. A. (Eds.), Psychosocial dynamics of cybersecurity (pp. 1–12). New York: Routledge.

Zheng, S., Yan, M., Liang, Y., Chen, Y., Wei, Q., & Li, S. (2024). Understanding the Positive and Negative Effects of Team Virtuality: A Theoretical Review and Research Agenda. Human Resource Management Review, 34(2). https://doi.org/10.1016/j.hrmr.2024.101013

Zhong, C., Alnusair, A., Sayger, B., Troxell, A., & Yao, J. (2019, April). AOH-map: A mind mapping system for supporting collaborative cyber security analysis. In *2019 IEEE Conference on Cognitive and Computational Aspects of Situation Management (CogSIMA)* (pp. 74–80). IEEE.

Zhou, Y., Cheung, C. M., & Hsu, S. (2017). A Dimensional Model for Describing and Differentiating Project Teams, International Journal of Project Management, 35(6), 1052–1065.

Zimmermann, V., & Renaud, K. (2019). Moving from a "Human-as-Problem" to a "Human-as-Solution" Cybersecurity Mindset. International Journal of Human-Computer Studies, 131, 169–187.

6 Oppositional Human Factors

Robert S. Gutzwiller, Christina M. Lewis, Rebecca Pharmer, Benjamin A. Clegg, Palvi Aggarwal, Giovanna Henery, Indrakshi Ray, and Rosa Martey

> A team of attackers are breaching into a defense network. They have a clear set of goals; conduct reconnaissance on the network, define its topology and then locate and exfiltrate content of interest – all without disrupting the network or alerting defenses. This is not their first challenge, and they suspect the network is heavily defended. They are also increasingly aware that there could be deceptive elements present on the network to disrupt or detect their actions. As they work through the layers of the network, a member on the team begins to notice that there are a handful of machines that appear vulnerable and would make good targets for exploitation. However, the machines also appear to have properties that make the attacker suspicious of them; the team begins to spend much of their time identifying any element they find as either real, or potentially fake (a honeypot or other deceptive component that should not be interacted with). Although not their primary task, it begins to dominate the limited time they have to achieve their mission.

The example above is a narrative summarization of one of the earliest available experiments conducted to study the behavior of cyber analysts in various situations involving cyber deception, both actual and psychological (Ferguson-Walter et al., 2017). The motivations for this work, and much of what follows, derives from remembering that unlike a "Mr. Robot" type of persona, cyber actors are still *human*. Thus, analysts are not just acting against systems – but are humans thinking about them and being influenced by their environment (such as a suggestion there may be deception; Ferguson-Walter et al., 2021). The observation that cyber actors themselves should be studied from a human factors and psychological viewpoint is a very new venture. This approach creates the opportunity to study related phenomena in a cybersecurity context.

The asymmetric disadvantage of defensive cyber operations serves as an obvious, important reason to study cyber operators. As of the 2025 Verizon Data Breach Investigations Report, several primary access methods of attackers are stolen credentials, phishing, and – importantly – the exploitation of vulnerabilities in the system or enterprise (Verizon, 2025). In the narrative above, it is easy to imagine a 'cyber-y' environment, complete with basement-dwelling, punk-ish protagonist anti-heroes, who enjoy wrecking their virtual havoc, crashing or infiltrating critical systems in the world. And while imagining those individuals, and perhaps even their

DOI: 10.1201/9781003413707-7

motivations is quite easy, understanding how the decisions are made in the attacking sequence of events is infinitely more difficult.

The continual effort toward improving defense of networks has taken many shapes (and continues to) but has largely been focused on the *technological* elements present in the system. Everything connected to elements in a network is quite available to potential attackers, even if large efforts have been made to hide them or bar entry. However, the reality revealed through many writings and investigations in cybersecurity operations, and particularly those of defenders, is that the work has many human and behavioral elements. The work is tedious and difficult (Zhong et al., 2019), overloading (D'Amico et al., 2005; Pfleeger & Caputo, 2012), fatiguing (Dykstra & Paul, 2018; Paul & Dykstra, 2017), boring, and plainly ill-suited to human capabilities (Gutzwiller et al., 2015; Sawyer et al., 2014). However, the human operator's cognition in this domain was generally understudied, and at a time when the US government was rapidly appreciating the scope of the cybersecurity problem looming on the horizon (Singer & Friedman, 2014). The creation and unification of cyber mission forces, and the stand-up of US Cyber Command, highlighted the important role of mixed teams for defense, or blue teams, and (though far less was said) offensive oriented "red" teams.

A major effort toward improving cyber defense involved understanding the nature of cyber analyst work. Around the same time, at a Human Factors and Ergonomics annual meeting, leaders from cybersecurity in the US Department of Defense even helped reach out to the research communities and explain some of the basic, human-centric challenges (Trent et al., 2016). In human factors and psychology, understanding what a human does in a complex task is routinely examined by conducting some type of task analytic work. Task analysis (generically) can take the form of cognitive interviewing, observations, tracking performance, and using various other methods to elicit knowledge from the people doing the work. The elicitation is usually about their goals, their methods, and the information they need to be successful in the processes. Combined this helps develop processes, displays, training, and many other facets of good human system integration (Endsley & Jones, 2012). A large number have been published in the open scientific literature (Champion et al., 2014; Clark et al., 2006; D'Amico et al., 2005; Erbacher et al., 2010; Gutzwiller et al., 2016a; Hoffman, 2005; Mahoney et al., 2010; Nyre-yu et al., 2019; Trent et al., 2019; Zhong et al., 2015), and have greatly increased the likelihood that cyber blue team tools, training, processes, and general work can be improved in careful, structured ways.

We generally avoided the literature on traditional users and the problem of scam or 'phish' emails and other security issues. In those works, users fail to follow security practices and maintain good cyber hygiene. While those individuals are absolutely a weak link in security – and addressing these issues will of course improve defense – they *are not cyber operators*. They have little to no training, and usually a job that does not deal directly with or focus on cybersecurity despite the *security* needs they have. Phishing research also uses tasks, e.g., users sifting through emails, or setting privacy settings, that do not have much to do with the work of network defense operations. Unfortunately, this meant that there was little to be gained from the extensive work. However, there *was* a commonality between user-based attempts to avoid phishing or scams, and cyber operators' daily work. The threats are constantly changing, and so are the defenses against them. From this realization sprung one enticing idea to help deal with the asymmetry (Gutzwiller et al., 2018): what if we could learn how to disrupt the attackers themselves, to supplement the technology *du jour*? No matter what defense you have, it is obsolete the instant a new possible vulnerability is discovered, methods that could disrupt an attacker's *mind* would be much harder to thwart. Disrupting attackers' thinking, their decision-making, and their choices as they encountered various aspects of a network: this was a potential solution that could scale across many people, require minimal technology to implement, and is based on deep psychological scientific perspectives. We ended up calling this theory *Oppositional Human Factors (OHF)*, as *any* usability or human factors advice on how to improve utility to a user should essentially be flipped or reversed for an attacker (Gutzwiller et al., 2018).

In the next sections, we present an overview of the current and future study of cyberattacker decision-making disruption as part of the OHF framework. A full review of cognitive aspects or other human factors practices that apply to cyber operations is beyond the scope of this chapter (but see Gutzwiller et al., 2015). These should always be evolving as we gain insight into human behavior and actions in cyberspace. Here, we focus mostly on decision-making and human attention, and what has been learned so far through the exciting investigations of realistic potential attackers, often using highly skilled red team participants.

COGNITIVE PSYCHOLOGY AND COGNITIVE BIASES

Human cognitive psychology is routinely defined by the study of the mind. Its origination out of the behaviorist theories of the time in the 1950s was marked by the notion that humans think and act, sometimes in ways that are counterintuitive, and that simple reinforcement or punishment schedules could not account for the general range of human behavior. As the cognitive branch of psychology grew, so too did the ability of scientists to measure and understand what would become foundational elements of thinking, including memory, attention, perception, action, and decision-making itself. It was during this period that several researchers can be identified that led to the current approach around biases in human reasoning and a modern scientific understanding of cognition. Prominent individuals such as Ulric Neisser, Donald Broadbent, and Don Norman highlighted the field of applied psychology by arguing that psychology needed to be examined 'outside the lab', and that its greatest value (and most interesting test of its utility and reliability) was likely to be discovered by studying the 'real world'. Departing the sterility of the laboratory but keeping rigorous theory, methods, and analytics intact has allowed the examination of these cyber behaviors. Further, this approach encouraged continuing theoretical development, a core portion of rigorous scientific work, and to which we believe the team has contributed.

While important early contributions to understanding human decision-making can be found in the work of Herb Simon, a pair of psychologists, Daniel Kahneman and Amos Tversky, offered what became a transformative approach. Their work, though it was often done using basic tests and students in classrooms, shed light on a growing and objective realization that humans were not completely rational actors. Humans could be swayed or misled even in logical scenarios. We use mental shortcuts (*heuristics*) that operate automatically and outside our conscious awareness to arrive at what Gigerenzer subsequently called 'fast and frugal' solutions (2008). However, these heuristics also leave us open to systematic errors – *cognitive biases*. Kahneman went on to be one of the few psychologists to be the recipient of a Nobel Prize (Ardila, 2016), and his bestselling book (Kahneman, 2011) propelled his life's work and others in the domain into public recognition. As a result, there has been increasing societal awareness and study of these biases (Pohl, 2022). Most of this has been good; the Obama administration, for example, recognized the biased way in which information was being presented to credit card holders that made them *more* likely to end up in critical debt, and enacted transparency legislation that dictated a better formatting of information. Popularization of biases can also have negative consequences; there is little doubt that these cognitive and attentional phenomena have long been exploited in casinos and gambling, occasionally form the backbone of dark design patterns for marketers (Nodder, 2013) including marketers like Amazon (O'brien & Katten, 2025), and strongly contribute to over-engagement in online social media (Hari, 2022).

The variety of existing biases (Johnson et al., 2020) that could be related to cybersecurity is a bit overwhelming, and it can be tempting to accept frames that group biases into various categories. The main Wikipedia entry for *cognitive bias* contains an infographic 'bias codex' which *purports* to organize these biases into major categories to explain their mechanism or origin. However, we found early on that this infographic was created by designers without cognitive backgrounds and grouped without any verification or validation by scientists. In other words, the Wikipedia infographic is not evidence-based but continues to mislead many people, even researchers(!), most likely when they are searching for helpful slide deck content. It should not be used to guide research or policy. Further, the infographic contained many so-called biases that were colloquial or anecdotal at best and had not been tested empirically or in experimental fashion, as well as attributes from other aspects of cognition that few decision scientists would regard as examples of cognitive biases because they do not involve the use of heuristics. Application of biases then cannot come simply from references to them or this kind of nice-looking high-level understanding: *especially* if it is a misunderstood and an inaccurate one.

Varying outcomes and rates of occurrence from biases have been found as well, with some biases such as confirmation bias (Nickerson, 1998; Wason, 1960) being pervasive, while others were more variable depending on conditions or situations such as sunk cost, for example, or the default effect. Some biases had few actual studies that revealed it, had only minor or inconsistent effects, or required very specific conditions to elicit (representativeness biases in general; see Gertner et al., 2016; Johnson et al., 2020).

Together, biases and cognitive psychological phenomena represent something knowable and unifying about cyber actors. Although it may be thought that cyberattackers are particularly unique or mentally robust, there is accumulating evidence (Gutzwiller et al., 2018; 2019; 2024) that they have standard cognition as it relates to bias (even if they may not think they do; see Ehrlinger et al., 2005; Pronin et al., 2002). While expertise makes people exceptional on some tasks, it can also create vulnerable situations where expectations are violated, or where a commonly useful assumption is actually wrong. We take the perspective from cognitive psychology that heuristics – decision-making shortcuts – are mostly beneficial but can be wrong in systematic ways (Gigerenzer, 2008; Kahneman & Klein, 2009). Since attackers are vulnerable to the same biasing errors discovered in most other human populations, one way to disrupt them would be to create situations in which biased behaviors are more likely. Even better would be scenarios in which the mistakes themselves result in behaviors that are more noticeable, able to be countered, and/or able to be mitigated by network defenders.

The study of biased decisions and later heuristics formed a major part of our OHF investigations. We began serious OHF efforts by examining over a hundred potential human cognitive and perceptual biases for potential fit to instances within cybersecurity operations, both defensive and offensive known to us (Johnson et al., 2020). Then we incorporated the opinion, feedback, and brainstorming of cybersecurity experts in both defense and offensive domains (Johnson et al., 2020, 2021) to find biases that appeared to be of the highest empirical value, based on potential effects and effect likelihood. The earliest lists prioritized examinations of the sunk cost fallacy (SCF), confirmation bias, framing effects, anchoring, and asymmetric dominance (AD; though other biases have been examined, such as default effect and illusory correlations).

THE STUDY OF OPPOSITIONAL HUMAN FACTORS

OHF for attacker decisions has been studied in two central ways. First has been examination of realistic cybersecurity experiments involving real qualified human attackers under observable conditions. The results from highly realistic studies are observational or exploratory, yet they are done with professional operators (Gutzwiller et al., 2024). This makes them crucial to understanding whether OHF approaches to security may work by indicating that the representative population is prone to biases in their area of expertise.

The second method comprises setting up experiments specifically to manipulate a set of defensive measures developed to elicit a bias or OHF effect and measuring the resulting behaviors. Researchers in these cases have conducted experiments across a variety of experimental platforms that aim to create various bias situations, whose effects are then evaluated. As these are new, and often complex, we offload the details to other manuscripts now and in the future and focus instead on showing how these biases can be investigated in meaningful cyber-like environments, offering evidence to the reader that this approach and its outcomes are useful.

EXAMINING AND EXPLORING BEHAVIOR FROM REALISTIC EXERCISES

There are numerous works that focus on understanding behavior from data collected in realistic cybersecurity experiments. Often these experiments consisted of attackers (often labeled 'red teamers') interacting with realistic systems, software, and infrastructure to accomplish a realistic mission. Red teams are groups of people "authorized and organized to emulate a potential adversary's attack or exploitation capabilities against an enterprise's security posture" ('Red Team' definition) (NIST, 2025). Two of these experiments contributed greatly to OHF and our initial forays into understanding the potential for biases in decision-making.

A dataset collected by Ferguson-Walter et al. (2017) was initially mined to determine whether chat reports and human behavior of red teamers could reveal any evidence of cognitive biases in action. Three publications explored the variety of different biases and behaviors that were found in that data set: Gutzwiller et al. (2018) established OHF; Gutzwiller et al. (2019) focused on some of the specific evidence that biases could be found in the data; and Johnson et al. (2022) explored the idea of team-level biases in cyber red teamers and developed a schema for team-level considerations of bias).

The most outstanding example of highly realistic data collected from professional red teamers who has contributed to the OHF methodology came from an experiment named Tularosa (Ferguson-Walter et al., 2019a; 2019b). Much of the detail can be found in the original primary source materials; for our purposes of description and relevance to this chapter, we point out the most impactful components: the experiment used more than 130 experts doing realistic activities over a two-day exercise carefully recorded. As a result, the dataset contained surveys, measures, and data collected including reporting and cyber performance components. The reporting requirements and chat clients were established so that red teamers could update, track progress, and report out things of importance as instructed. The major outcomes of this experiment have been spread across a few publications (Ferguson-Walter et al., 2019a/2019b; 2021; 2023) and have focused on the efficacy of using both cyber deception and psychological deception during cyber defense. However, as a major contribution to studying OHF related to biases, Gutzwiller et al. (2024) published a rigorous examination of the dataset for biases and found strong evidence for both confirmation bias and framing effects. Two further papers using this dataset also explored and found various emotionality and affective components of behavior that were related to disrupting attackers using various defensive techniques (Ferguson-Walter et al., 2021; Gutzwiller et al., 2023).

Despite the large amount of evidence gathered for OHF, neither of the experiments above were initially designed specifically to test any specific decision-making bias implementation of the OHF concept. To do that, our broad team worked to design and run specific tests of biases in cyber-relevant paradigms that we could control, many of which we had to create from scratch with the help of experts.

CONDUCTING RIGOROUS OHF EXPERIMENTS

In this section, we summarize key experiments designed to test hypotheses and manipulations of decision-making biases and other OHF techniques. While a full discussion of all projects and biases is out of scope here, we have summarized some of the key findings to showcase some of these initial experiments and results.

Sunk Cost Fallacy

SCF (Arkes & Ayton, 1999; Arkes & Blumer, 1985; Arkes & Hutzel, 2000; Staw, 1976) can be viewed as an example of loss aversion; when given a choice, people prefer to avoid certain losses. As a result, once a person has made prior investments, such as money, effort, or time into a task, they also avoid discontinuing that task (Arkes & Blumer, 1985), as deviating from that course of action is perceived to be a 'loss' of those investments. Cyberattackers who invest a lot of resources in an attack process, for example, may tend to incorporate these historical costs in assessing the value of *continuing* a course of action. This is potentially the case even when they are given options that have a chance of being more successful, even though costs already spent should carry less or no weight in decisions about current or incremental investments for the future (Thaler, 1980).

A variety of research on SCF exists in human subjects' experiments, and ranges from studies of economics, business and finance, marketing, sports, tourism, consumerism, and decision-making. According to a meta-analysis by Roth et al. (2015), SCF research falls into two main decision formats: *utilization decisions* and *progress decisions.*

Utilization decisions around sunk cost describe "a decision-maker confronted with the choice between two equally attractive alternatives, such that preferences shift to the sunk-cost alternative" (Roth et al., 2015, p. 100). In other words, the SCF, when recorded in these experiments, relies on people selecting an option that continues the plan and investment rather than an alternative. These tend to be single decisions, wherein participants are presented with a fictitious situation or narrative that creates a 'sunk cost' investment and then confronts the participant with a choice. Consider this example, from experiment 3 in Arkes and Blumer (1985):

> As the president of an airline company, you have invested 10 million dollars of the company's money into a research project. The purpose was to build a plane that would not be detected by conventional radar, in other words, a radar-blank plane. When the project is 90% completed, another firm begins marketing a plane that cannot be detected by radar. Also, it is apparent that their plane is much faster and far more economical than the plane your company is building. The question is: should you invest the last 10% of the research funds to finish your plane?.
>
> **p. 129**

Most participants given this question say 'yes' (41/48 in their experiment), which means an additional $1M may be spent on a failing idea. When given similar information, but *without* the 'sunk cost' of having spent money already, participants taking the same perspective of president choose *not* to invest the last $1M (50/60). In other words, a history of investment may sway decision-making in significant and detrimental ways, if continuing it could be costly. Such history can be *provided* (as in the written example) or *experienced* (by a real person, investing time, money, or energy into a particular course of action or path).

We take the approach of given participants experienced history and making more iterative decisions over time, which we believe to be more reflective of realistic world conditions. We also rely on the second sunk cost decision format of progress decision research in sunk cost. Progress decisions focus on assessing *multiple* decisions made over an entire course of action. For example, Staw et al. (1997) investigated how banks dealt with the accumulation of problematic loans. They analyzed decisions about bad debt over nine years across 132 banks and found that managers who had been with the bank for a longer time tended to hold on to failing loans longer (Staw et al., 1997). Thus, this SCF in decision-making over time greatly increased institutional losses. Like in progress decisions (Kanodia et al., 1989), in cybersecurity, decisions are chained together over some period.

We believe this same type of decision-making occurs in cybersecurity during attack phases. Attackers invest time, money, and mental energy into some courses of action, which influence their later decisions about whether to continue or withdraw (e.g., resources spent attempting to get root access/permissions to a machine but later discovering a potentially easier method to access a network). Further, if those decisions are affected by sunk cost, then the amount of time/money/energy may weigh them to continue their plan versus choose a different (potentially better) plan. For now, we confined our investigation to individuals but believe this could happen in larger groups who divvy tasks out to different members of a team.

While there is evidence that the SCF can be induced in various situations, we remain unaware of any other research aside from our own that empirically investigates this decision-making bias within a cyber environment or in an attack kill chain context.

METHOD

Some of the earliest cyber-oriented experiments around sunk cost were developed (as outlined in Johnson et al. (2024)) and run using Mechanical Turk recruited participants. These studies utilized a code-breaking game we helped develop called Cypher in which participants tried to break codes using a cypher key (Figure 6.1).

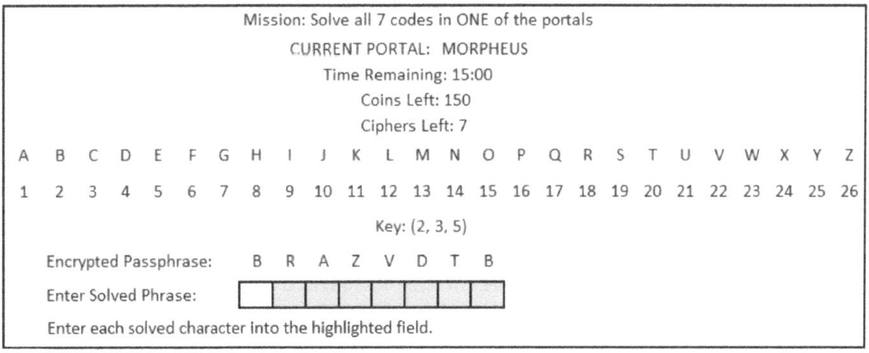

FIGURE 6.1 Cypher game interface, solving a single cypher. Participants were able to use the key presented along with the alphabet to solve the encrypted phrase at the bottom of the screen. Incorrect responses in the boxes triggered a slight delay, as well as a penalty, to encourage accurate responding and made sure participants would avoid spamming keys. Trials were time limited.

Codes had various difficulty and lengths, and participants were given options in the course of solving them – in the form of a choice about what sets of those were available to work on – and they could choose to continue cracking their current set, or switch to a different set with different difficulty, etc. This comprised the core of the sunk cost measure, as we were able to see whether a participant ‘stayed’ or ‘switched’ based on their progress through the game.

Additionally, we manipulated the certainty (or uncertainty) of the message given at the switch opportunity, as well as when the switch itself was offered (earlier or later) as well as the overall difficulty of the cyphers (easier or harder to solve). The full details of the studies and results can be found in Johnson et al. (2024). These initial experiments supported that SCF could be elicited in this paradigm, and was influenced by uncertainty as well as project completion (fewer switches taken when the switch opportunity was presented later rather than earlier), although not always as predicted. Generally, participants stayed more often than they switched when given the opportunity (Johnson et al., 2024).

Together, both Cypher experiments revealed key evidence toward exploring the use of SCF for biasing attackers. It is worth pointing out that many investigations of SCF up to this point were utilization decisions, often conveyed in brief paragraphs to participants and unmoored from any practical reality. Our experiment went well beyond this approach, and involved conceptual progress decisions, in which participants had to decide whether to continue, and in an environment where they literally had a stake in the outcome of that choice.

Asymmetric Dominance and Anchoring Biases

Results from experiments attempting OHF using other biases, such as AD, have also been promising but not always as clear. The AD effect comes from marketing situations in which people are faced with a decision about what product to purchase (shoes, for example), with each product varying across multiple properties that matter to them (price and comfort). Researchers then manipulate the properties of the items to create a two-choice comparison (shoe A is $59 but low comfort; shoe B is $125 and high comfort). The utility of the AD effect is that one can change the purchase decision between A and B by simply adding a third option in that is clearly worse on one attribute (shoe C, the ‘decoy’ shoe, which is $125, but slightly less comfortable than shoe B). In these kinds of situations, the presence of the decoy will alter the percentage of people choosing the high cost and high comfort shoe (see Ariely & Wallsten, 1995).

A second experimental platform, attack surface (AS), was developed for measuring decision-making biases. Building on lessons learned in the Cypher experiments, AS presented an abstracted networks of personal computers and servers where the task was to attack systems and gain points en route to discovering a valuable data token. We took a similar approach described in the AD literature and designed experiment scenarios in which an attacker would encounter two (or three, in the AD condition) servers on a network with varying properties (cost and number of successful attacks required). We kept in mind a host of relevant details regarding effective creation of the bias, such as ensuring the set of choices and attributes was perceived in full (e.g., Simonson, 2014), providing a relatively low number of items in the choice set (Stanley & Wedell, 2023), ensuring both initial options were approximately equal (Malkoc et al., 2013; Pettibone, 2012) and establishing meaningful attributes for each option (Padamwar et al., 2023).

Using the AS platform, the AD effect was tested by manipulating the presented costs of servers (hosts) for each attack, and the number of successful attacks required to fully compromise the system at a specific location in each network. Within each of 12 network trials in this study, the test point was presented with either two or three servers to attack, depending on the between-subjects condition: a control (two-choice) condition, with an expensive, target host 'X' (which we want attackers to select as it may waste their time and resources), and a much proportionally cheaper host 'Z'. Or, for the experiment AD (three-choice) condition, we included the same 'X' and 'Z' choices along with a third server 'Y' that was asymmetrically dominated by the target choice 'X'. Here, the coin cost for 'Y' was only slightly cheaper than 'X' but had additional required attacks. The AD literature predicts that people would select the more expensive option 'X' more often under these conditions than in control.

These test points were encountered as part of normal AS gameplay but were placed at different levels across the trials (either in subnet 2 or 3, or in subnet 4 or 5). Participants were instructed to try and reach their point goal garnered through successfully attacking systems for each trial before running out of coins, turns, or time. The presentation order of each system in a subnet was randomized. Thus, our participants could have seen the set as ['Y' 'Z' 'X'] or ['Z' 'X' 'Y'], etc. Immediate feedback about the success or failure of each attack, coins expended, and points gained was also provided.

Our preliminary analysis of choice showed that across all trials, the choice *was* influenced by the presence of the AD machine; however, the direction of the effect was the opposite of expectations; the addition of the AD item 'Y' was supposed to drive behavior toward the expensive option, instead, the cheaper item was more likely to be initially attacked in the AD condition. The stage of the game also had a significant effect: during the first half of the game, participants were more likely to first attack the cheap 'Z' host, and it appeared that over time participants began to eventually switch their pattern and first attacked the 'X' more often. However, the proportion of selecting 'Z' was still higher for the AD experimental condition.

The decoy itself appeared to be successful (in AD experiments, participants should rarely ever select a decoy option, which we found was the case). However, we speculated that there seemed to be a learning effect where participants became more likely to attack the expensive choice as the game progressed, as they may have developed an expectation that high-cost servers garnered the highest point returns. Finally, in most AD effect investigations in the research literature, participants receive no feedback on their choice selection. In *contrast*, our feedback, which is akin to a real cyber scenario, could have partially contributed to an observed difference between the choice of server for participants' first attacks, and the first full compromise of a server. In other words, participants receiving feedback after the first click may have changed their minds in the face of new information.

Anchoring Bias Definition

In everyday situations, people making decisions can show a tendency to be highly influenced by the initial piece of information they are given or think of, and they fail to update their thinking and understanding as new information becomes available.

Everyday example:	**Cyber example:**
For example, if asked about the length of time an African elephant is pregnant before giving birth, most people start from the human equivalent of 9 months and add a few on to that in their guess but then fall far short of the actual answer of 22 months.	For example, if an attacker is using credential stuffing attacks, the attacker will assume that the login information that they have can be used on multiple websites and will test for that.

FIGURE 6.2 Anchoring bias.

Additional experiments used other experimental platforms and sometimes combined the implementation of biases. Anchoring and adjustment is a cognitive bias that reflects a process whereby people tend to be overly influenced by an initial piece of information in their subsequent decision (see Tversky & Kahneman, 1974). As we reach decisions, human decision-makers tend to begin from a starting point that might be internally generated or externally supplied. They then make adjustments to their estimate from that initial information until they arrive at what seems a plausible answer, although often with insufficient movement from their starting point. The initial information thus serves as an anchor for their answer, with low anchors producing lower final determinations, whereas high anchors tend toward higher final determinations.

Pharmer et al. (2024) tested OHF ideas surrounding the use of anchoring effects (studying the challenge to using anchoring in the applied world, the authors examined both 'global' anchors and more locally encountered anchors simultaneously; this is a common challenge of application versus more laboratory-based experimentations of anchoring and adjustment).

In this experiment, participants took the role of a white hat actor attempting to regain stolen money from a black hat hacker's bank. The paradigm required multiple types of decisions, embedded within a flow of other activities – moving beyond many of the simplistic, one-shot decisions classically used to examine cognitive biases (Figure 6.2). The game allowed for ten rounds of play, during which participants would see various bank accounts along with the funds in them, and were given a choice to either 'steal' that amount (which would elevate the risk of being caught by 5%, although participants were not informed of the actual amount of change in their risk) or not steal and move to another account (which also increased risk over time but at a 1% rate). Being caught resulted in a complete loss of all money for a given round, while surviving to the last trial resulted in successfully retrieving the funds. A global anchor was explored through instructions for the task, which varied between informing participants that either 'stealing from 3 or more accounts' or 'stealing money from too many accounts' had a very high risk of detection. A more traditional local anchor was provided by varying the first piece of information encountered; in this case, the initial account dollar value of each block of trials with a high value (~$215k), low value (~$5k), or a random value intended to function as a control (average $110k, in a range from $20k to $200k).

Pharmer et al. (2024) results showed that the global anchor did not produce an effect overall on total times caught, and neither did the local account anchoring on the value of accounts chosen to attack (except in conditions without the global anchor). However, when limiting the analysis to the first block of trials, or round of play, both global and local anchors had their intended effects; they were not found when examining the second block, suggesting their decay is rapid. This has obvious implications for cyber contexts in which the establishment of an anchor – whether it is local or global – will need to be salient and may have limited duration of effect. When an attacker becomes anchored then, they may have to be *immediately* 'steered' toward a defense-chosen destination or concept (such as a particular honeypot).

Second, Pharmer et al. (2024) explored the AD effect by creating sets of vulnerable machines to attack, which participants selected among before starting the account task. The machines varied across two properties (operating system and security vulnerabilities) with the idea that more vulnerabilities could indicate a more successful machine to attack, but that older operating system (which may have more vulnerabilities) are also more likely to be deceptive honeypot machines that can catch attackers. The AD conditions compared two versions of this choice: one in which two machines were offered (Win 10 with moderate vulnerabilities; Win 11 with minimal vulnerabilities), and a second set in which a third machine was offered (Win 7 with some vulnerabilities), designed to push attackers toward attacking a high-risk machine. The asymmetric dominant alternative generally failed to steer people away from the most secure system as intended. However, when narrowing the analysis only to those participants who had been in the global anchoring condition (which specified, 'stealing money from three or more accounts' using the language as explained above), there was some evidence of a dominance effect, with a moderate change from 57% selection of the high-risk system without the decoy to 35% selection with the decoy. The authors speculate that this difference is a result of reduced *attention* paid by the non-domain expert participants to details (a general requirement for AD effects to work), which could have been brought on by the vague language used in the anchoring condition. Overall, the findings offer promise for the use of cognitive biases in cyber contexts but with several critical caveats. One important note is that the incorporation of multiple, concurrent biases tended to negate the effects observed. This suggests implementation needs to be far more than undirected, 'shotgun' approaches that seek to scatter potential bias inducing elements throughout a system. Taken together, the results from the above experiments also highlight the critical role of attention in the success of OHF manipulations.

OHF through Attention Disruption and Distraction

At the outset of OHF, it was identified that many aspects of a human operator could be compromising to attackers (Gutzwiller et al., 2018). Human attention is limited (Grier et al., 2008; Wickens, 2002; 2008; 2024) and one that must be allocated well to perform complex tasks (Wickens et al., 2022). This is especially true when time is limited, the users are stressed or fatigued, or the work is particularly difficult and thus approaching a user's capacity (e.g., 'redline'). Phenomena that are present in these conditions are familiar to most people as distraction – a lack of focus or a focus on the wrong things, as well as attentional tunneling (overly focused behavior on something that may still be important but at the expense of something else important) and the difficulty in determining among alternatives, when many are present and time limited, what task to do (Barg-Walkow et al., 2021; Gutzwiller et al., 2016b; 2019; Lewis et al., 2024; Wickens & Gutzwiller, 2017; Wickens et al., 2015; 2016).

So far, OHF has not attempted to manipulate attacker attention or utilize these limitations to defenders' advantage, although evidence of attention's influence has been found in the form of distraction from goals (Gutzwiller et al., 2018). In that work, we pointed out, for example, a decoy system or honeypot's purpose might be to allow an attacker to begin downloading a large seemingly valuable file but then to degrade or to cease this file transfer as soon as the attacker begins interacting with other (more valuable) information systems. The uncertainty and opportunity costs associated with initiating interaction with a new, unknown system – versus the 'known' value of the existing file transfer – is likely to lead the attacker into fixating their attention (and time, and resources) into the task of determining the reason for the file transfer failure. This is a reoccurring facet of human attention and can be found to be part of root cause in aviation incidents and accidents; our goal would be to create these types of events and situations such that attention is diverted, and as a result, the defenders get an advantage.

A study is in progress that captures the concept of OHF to disrupt, interrupt, or even prevent attack success by utilizing cyber and psychological deceptions on the attacker's attention. Attention is the state of consciousness in which a person responds to a stimulus or stimuli. Although many aspects of attention have been studied in psychology, there is insufficient research conducted in the field of cybersecurity, not only in its theory but also in its application. Thus, attention theory can be leveraged to develop robust defenses against potential attackers. However, it is difficult to assess the factors of attention as human attention mechanisms are complex. Cybersecurity can employ attention theory in two ways:

- diverting the attacker's attention to tasks that do not contribute to a successful attack of a system or,
- influence the attackers to home in on tasks that fail to undermine the defense of the system.

The task being performed by the attacker can be classified as ongoing task (OT) and the tasks which can be switched to by attackers are classified as alternate tasks (AT). This study will aim to predominantly utilize the attentional factors of saliency and difficulty to increase the 'attractiveness', diversion of attention to AT, or 'stickiness', focus attention on the OT (Figure 6.3).

The strategic task overload management (STOM) model (e.g., Wickens et al., 2015) evaluates attentional factors such as saliency, interest, difficulty, and priority to understand the nature of task switching among attackers. In this project, we are using manipulations of some of these attentional factors to better understand their contribution in influencing the decision-making of attackers (specifically saliency and difficulty). Both saliency and difficulty are manipulated in a cybersecurity task where attackers collect information about various nodes in the network (network

Factors for Attention	Explanation
Salience	Ability of the arrival of a task to 'call attention to itself' / visual distinctiveness of an object
Interest	Engaging ability of task
Priority	Relative importance of a task
Difficulty	Mental workload imposed by a task
Automation reliability	Trust and dependence on automation resources
Attractiveness	The ability of an alternative task to draw attention to itself
Stickiness	The ability of the ongoing task to keep attention to itself

FIGURE 6.3 Attention factors explored in experimental design for OHF experimentation.

reconnaissance phase), exploit the systems (exploitation phase), and perform data exfiltration by either cracking the passwords or guessing the passwords (data exfiltration phase). During the reconnaissance phase, participants conduct an Nmap scan on a targeted system, which outputs a vulnerability table that includes exploitable ports, vulnerability names common weakness enumeration (CWE) IDs, and the likelihood of a successful exploit. The salience of information is manipulated through two variables: color and likelihood. In the 'Neutral' condition, the table is devoid of color, whereas in the 'Minimum' condition, vulnerabilities are highlighted in yellow. Regarding likelihood, it is tested between subjects the 'Random' condition involves arbitrary color assignments to vulnerabilities, and the 'Congruent' condition assigns yellow to vulnerabilities when the likelihood of success is high. Thus, each participant will encounter two 'Neutral' condition rounds and two 'Minimum' color conditions where both rounds may be 'Random' or 'Congruent' in its color assignment.

The exploitation phase follows, where participants select an exploit based on the information from the vulnerability table and initiate an attack on the system.

In the action on objectives phase, participants are tasked with either resetting or cracking passwords. The 'Choice' and 'No Choice' conditions are tested between subjects such that each participant is assigned one of the conditions for all four rounds with varying levels of difficulty in their tasks. In the 'No Choice' condition, participants are assigned a specific task – either a password reset or a password crack – with a designated level of difficulty (either hard or easy) for each system in each of the rounds with the option to decline the task and attack a different system. Conversely, in the 'Choice' condition, participants are presented with both tasks on each system in each of the rounds with varying difficulty in each of the tasks, allowing them the flexibility to attempt one, both, or neither of the tasks, or to log out and engage a new system.

This structured experimental design enables a detailed analysis of how variations in information salience and task difficulty influence strategic decision-making within a cybersecurity framework. While results are forthcoming, the project represents a novel expansion into additional behavioral disruptions and remains promising.

DISCUSSION

Far more exploration can be pursued in the OHF strategy. While collecting data in any human subjects' experiment is challenging and has limited the rate of exploration, the number of intriguing ideas and realism around OHF is increasing at a rapid pace. Disrupting elements of how attackers are thinking, attending, and deciding about their adversarial actions continues to have major potential.

Impact of Oppositional Human Factors

One of the biggest contributions from OHF has been a focus on the thinking, behavior, and human components of cyber operators. Across the current work, techniques have discovered impacts on emotionality (influenced confusion, surprise, and frustration) as well as decisions (creating confirmation bias, and opportunities to fall for sunk cost). The idea of disrupting these things as a defensive mechanism has largely been validated, though more evidence continues to be collected.

While not about decision-making per se, affective elements of cyber operators remain important as well. The psychological literature is less clear about the influence of emotionality on some of the standard cognitive biases, so one challenge here is having a clear prediction of expected outcomes. Similarly, while there are individual differences in cognitive biases, it is unclear whether these stem from a unified factor (e.g., Teovanović et al., 2015). Much of the bias work, in particular, is very sparse on mediator and moderator effects assessment. Other individual factors are also understudied (though there is limited work on the fatigue and burnout issues that are present in some levels of defense operations).

In sum, cognitive biases appear in real, trained cyber operators under realistic conditions where they can be observed. This requires careful examination, careful effort, and educated analysis. As previously mentioned, a 'shotgun' approach may not work. The upside is that methods for bias scoring may be applicable to previously collected data, and there is also the potential to use machine learning or other methods to discover how to automatically classify this behavior and potentially the susceptibility to OHF techniques to allow for more tailored approaches.

Continuing Challenges to OHF and Disrupting Attackers

Biases seem to be manipulated through experimental protocols but not yet in a reliable manner. On the one hand, this speaks to the 'applied' component of these efforts, and on the other is also likely attributed to the nature of some of the biases themselves. As mentioned, biases like sunk cost are not routinely tested in more realistic settings which makes doing so an additional challenge. Nevertheless, we still found consistent evidence for them.

While much of the chapter has focused on the OHF concept, so much remains to be done with humans in cybersecurity if it is to be validated and applied more broadly. Studying people doing their job is at the heart of good operational science in the human factors and ergonomics communities, but some older detailed task-analytic work is aging as the pace of change in operations, processes, and technology is rapid. The more one understands the potential adversary, the better and more reliable a response to mitigate their actions should be, combining a study of the psychological elements at play in cyber with the technical aims to improve the system's security.

Biases or methods of OHF themselves can be pervasive, and ready to apply to complex topics like cyber operations. They show up in most people, even in the face of experience and high intelligence in many cases. The excitement is in matching the OHF technique to a realistic scenario configuration, and being able to impose the bias architecture on tasks that attackers may need to do for their missions. This process is non-trivial, requires transdisciplinary teams of psychologists, engineers, and cyber analysts, and – having created something that might fit the bill – still requires validation in a realistic environment.

Finally, work in human factors suggests being cautious in thinking that artificial intelligence solutions and automation will solve these problems without creating new, sometimes worse challenges and consequences (Bainbridge, 1983; Strauch, 2017), even for modern cyber operations (Gutzwiller & Van Bruggen, 2021). For example, building more automation into enabling rapid defense responses can backfire by hiding how system management is handled from IT and CERT teams who need to know. However, automation could aid in managing and helping to implement OHF techniques once they are fully understood; it will also create a need to revisit the basics of how to integrate human operators with automated systems. This continues to suggest a broad need for experts to 'bridge' roles and merge or share knowledge between the psychology of human factors, cybersecurity operations, and computer science.

REFERENCES

Ardila, R. (2016). Psychology and the Nobel Prize. *Universitas Psychologica*, *15*(4), 1–6.

Ariely, D., & Wallsten, T. S. (1995). Seeking Subjective Dominance in Multidimensional Space: An Explanation of the Asymmetric Dominance Effect. *Organizational Behavior and Human Decision Processes*, *63*(3), 223–232.

Arkes, H., & Blumer, C. (1985). The Psychology of Sunk Cost. *Organizational Behavior and Human Decision Processes*, *35*, 124–140.

Arkes, H. R., & Ayton, P. (1999). The Sunk Cost and Concorde Effects: Are Humans Less Rational than Lower Animals? *Psychological Bulletin*, *125*(5), 591–600. https://doi.org/10.1037/0033-2909.125.5.591

Arkes, H. R., & Hutzel, L. (2000). The Role of Probability of Success Estimates in the Sunk Cost Effect. *Journal of Behavioral Decision Making*, *13*(3), 295–306. https://doi.org/10.1002/1099-0771(200007/09)13:3<295::AID-BDM353>3.0.CO;2-6

Bainbridge, L. (1983). Ironies of Automation. *Automatica*, *19*(6), 775–779.

Barg-Walkow, L. H., Thomas, R. P., Wickens, C. D., & Rogers, W. A. (2021). Modeling Task Scheduling Decisions of Emergency Department Physicians. *Human Factors*, *63*(3), 450–461. https://doi.org/10.1177/0018720819893427

Champion, M., Jariwala, S., Ward, P., & Cooke, N. J. (2014). Using Cognitive Task Analysis to Investigate the Contribution of Informal Education to Developing Cyber Security Expertise. *Proceedings of the Human Factors and Ergonomics Society Annual Meeting*, *58*, 310–314. https://doi.org/10.1177/1541931214581064

Clark, R. E., Feldon, D., van Merriënboer, J. J. G., Yates, K., and Early, S. (2008). Cognitive task analysis. In J. M. Spector, M. D. Merrill, J. J. G. van Merriënboer, & M. P. Driscoll (Eds.). Handbook of research on educational communications and technology (3rd ed.) (pp. 577-593). New York: Macmillan/Gale.

D'Amico, A. D., Whitley, K., Tesone, D., O'Brien, B., & Roth, E. (2005). Achieving Cyber Defense Situational Awareness: A Cognitive Task Analysis of Information Assurance

Analysts. *Proceedings of the Human Factors and Ergonomics Society Annual Meeting, 49*(3), 229–233. https://doi.org/10.1177/154193120504900304

Dykstra, J., & Paul, C. L. (2018). Cyber Operations Stress Survey (COSS): Studying Fatigue, Frustration, and Cognitive Workload in Cybersecurity Operations. USENIX Workshop on Cyber Security Experimentation and Test. https://www.usenix.org/conference/cset18/presentation/dykstra

Ehrlinger, J., Gilovich, T., & Ross, L. (2005). Peering into the Bias Blind Spot: People's Assessments of Bias in Themselves and Others. *Personality and Social Psychology Bulletin, 31*(5), 680–692. https://doi.org/10.1177/0146167204271570

Endsley, M. R., & Jones, D. (2012). *Designing for Situation Awareness: An Approach to Human-Centered Design* (2nd ed.). CRC Press.

Erbacher, R. F., Frincke, D. A., Wong, P. C., Moody, S., & Fink, G. (2010). Cognitive task analysis of network analysts and managers for network situational awareness. *Visualization and Data Analysis 2010, International Society for Optics and Photonics. 7530*(75300H), 12. https://doi.org/10.1117/12.845488

Ferguson-Walter, K. J., LaFon, D. S., & Shade, T. B. (2017). Friend or Faux: Deception for Cyber Defense. *Journal of Information Warfare, 16*(2), 28–42.

Ferguson-Walter, K. J., Major, M. M., Johnson, C., & Muhleman, D. (2021). Examining the Efficacy of Decoy-Based and Psychological Cyber Deception. *USENIX Security Symposium.* https://www.usenix.org/conference/usenixsecurity21/presentation/ferguson-walter

Ferguson-Walter, K. J., Major, M. M., Johnson, C. K., Johnson, C. J., Scott, D. D., Gutzwiller, R. S., & Shade, T. (2023). Cyber Expert Feedback: Experiences, Expectations, and Opinions about Cyber Deception. *Computers and Security, 130*(June). https://doi.org/10.1016/j.cose.2023.103268

Ferguson-Walter, K. J., Shade, T. B., Rogers, A. V., Niedbala, E., Trumbo, M., Nauer, K., Divis, K., Jones, A., Combs, A., & Abbott, R. (2019a). *Appendix* to The Tularosa Study: An Experimental Design and Implementation to Quantify the Effectiveness of Cyber Deception. https://www.researchgate.net/profile/Kimberly-Ferguson-Walter/publication/372160277_Appendix_to_The_Tularosa_Study_An_Experimental_Design_and_Implementation_to_Quantify_the_Effectiveness_of_Cyber_Deception/links/64a74489b9ed6874a5fff5be/Appendix-to-The-Tularosa-Study-An-Experimental-Design-and-Implementation-to-Quantify-the-Effectiveness-of-Cyber-Deception.pdf

Ferguson-Walter, K. J., Shade, T. B., Rogers, A. V., Niedbala, E., Trumbo, M., Nauer, K., Divis, K., Jones, A., Combs, A., & Abbott, R. (2019b). The Tularosa Study: An Experimental Design and Implementation to Quantify the Effectiveness of Cyber Deception. *52nd IEEE Hawaii International Conference on Systems Science (HICSS).* 7272–7281. https://hdl.handle.net/10125/60164

Gertner, A., Zaromb, F., Schneider, R., Roberts, R. D., & Matthews, G. (2016). The Assessment of Biases in Cognition. *MITRE Technical Report (MTR160163)*, 1–245. https://www.mitre.org/publications/technical-papers/the-assessment-of-biases-in-cognition

Gigerenzer, G. (2008). Why Heuristics Work. *Perspectives on Psychological Science, 3*(1), 20–29. https://doi.org/10.1111/j.1745-6916.2008.00058.x

Grier, R., Wickens, C., Kaber, D., Strayer, D., Boehm-Davis, D., Trafton, J., & St. John, M. (2008). The Red-Line of Workload: Theory, Research, and Design. *Proceedings of the Human Factors and Ergonomics Society Annual Meeting, 52*, 1204–1208. https://doi.org/10.1037/e578252012-011

Gutzwiller, R., Ferguson-walter, K., Fugate, S., & Rogers, A. (2018, September). "Oh, look, a butterfly!" A framework for distracting attackers to improve cyber defense. In *Proceedings of the Human Factors and Ergonomics Society* Annual Meeting (Vol. 62, No. 1, pp. 272–276). Sage CA: Los Angeles, CA: SAGE Publications.

Gutzwiller, R. S., Fugate, S., Sawyer, B. D., & Hancock, P. A. (2015). The Human Factors of Cyber Network Defense. *Proceedings of the Human Factors and Ergonomics Society Annual Meeting, 59*(1), 322–326. https://doi.org/10.1177/1541931215591067

Gutzwiller, R. S., Gilbert, M., Drescher, T. J., Ferguson-Walter, K. J., Mikanda, N., Johnson, C. J., & Scott, D. D. (2023, September). Frustration, Confusion, Surprise, Confidence, And Self-Doubt: Cyber Operators' Affects During A Realistic Experiment. In Proceedings of the Human Factors and Ergonomics Society Annual Meeting (Vol. 67, No. 1, pp. 233-239). Sage CA: Los Angeles, CA: SAGE Publications.

Gutzwiller, R. S., Hunt, S. M., & Lange, D. S. (2016a). A Task Analysis toward Characterizing Cyber-Cognitive Situation Awareness (CCSA) in Cyber Defense Analysts. *2016 IEEE International Multi-Disciplinary Conference on Cognitive Methods in Situation Awareness and Decision Support (CogSIMA)*, 14–20. https://doi.org/10.1109/COGSIMA.2016.7497780

Gutzwiller, R. S., Rheem, H., Ferguson-Walter, K. J., Lewis, C. M., Johnson, C. K., & Major, M. (2024). Exploratory Analysis of Decision-Making Biases of Professional Red Teamers in a Cyber-Attack Dataset. *Journal of Cognitive Engineering and Decision Making*, *18*(1), 37–51. https://doi.org/10.1177/15553434231217787

Gutzwiller, R. S., & Van Bruggen, D. (2021). Human Factors in Automating Cyber Operations. *Proceedings of the 54th Hawaii International Conference on System Sciences*, 1968–1978. https://doi.org/10.24251/HICSS.2021.241

Gutzwiller, R. S., Wickens, C. D., & Clegg, B. A. (2016b). The Role of Time on Task in Multi-Task Management. *Journal of Applied Research in Memory and Cognition*, *5*, 176–184. https://doi.org/10.1016/j.jarmac.2016.04.003

Gutzwiller, R. S., Wickens, C. D., & Clegg, B. A. (2019). The Role of Reward and Effort over Time in Task Switching. *Theoretical Issues in Ergonomics Science*, *20*(2), 196–214. https://doi.org/10.1080/1463922X.2018.1522556

Hari, J. (2022). *Stolen Focus: Why you can't pay attention--and how to think deeply again*. Crown.

Hoffman, R. R. (2005). Protocols for Cognitive Task Analysis. *Institute for Human and Machine Cognition*. http://ihmc.us:16080/research/projects/CTAProtocols

Johnson, C. J., Ferguson-walter, K. J., Gutzwiller, R. S., Scott, D. D., & Cooke, N. J. (2022). Investigating Cyber Attacker Team Cognition. *Proceedings of the Human Factors and Ergonomics Society*, *66*, 105–109.

Johnson, C. K., Gutzwiller, R. S., Ferguson-Walter, K. J., & Fugate, S. J. (2020). *A Cyber-Relevant Table of Decision Making Biases and Their Definitions*. https://www.researchgate.net/publication/344106644_A_Cyber-Relevant_Table_of_Decision_Making_Biases_and_their_Definitions

Johnson, C. K., Gutzwiller, R. S., Gervais, J., & Ferguson-Walter, K. J. (2021). Decision-Making Biases and Cyber Attackers. *Proceedings of the 36th IEEE/ACM International Conference on Automated Software Engineering Workshops*, 140–144. https://doi.org/10.1109/ASEW52652.2021.00038

Johnson, C. K., Van Tassel, R. W., Rogers, A., Shade, T., & Ferguson-Walter, K. (2024). Adversarial Cognitive Engineering (ACE) and Defensive Cybersecurity: Leveraging Attacker Decision-Making Heuristics in a Cybersecurity Task. *Proceedings of the Annual Hawaii International Conference on System Sciences*, *1*, 974–983.

Kahneman, D. (2011). *Thinking fast and slow*. Farrar, Strauss & Giroux.

Kahneman, D., & Klein, G. (2009). Conditions for Intuitive Expertise: A Failure to Disagree. *American Psychologist*, *64*(6), 515–526. https://doi.org/10.1037/a0016755

Kanodia, C., Bushman, R., & Dickhaut, J. (1989). Escalation Errors and the Sunk Cost Effect: An Explanation Based on Reputation and Information Asymmetries. *Journal of Accounting Research*, *27*(1), 59–77.

Lewis, C. M., Gutzwiller, R. S., & Johnson, C. K. (2024). Priority Influences Task Selection Decisions in Multi-Task Management. *Applied Ergonomics*, *119*, 104317. https://doi.org/10.1016/j.apergo.2024.104317

Mahoney, S., Roth, E., Steinke, K., Pfautz, J., Wu, C., & Farry, M. (2010). A Cognitive Task Analysis for Cyber Situational Awareness. *Proceedings of the Human Factors and Ergonomics Society Annual Meeting*, *54*, 279–283.

Malkoc, S. A., Hedgcock, W., & Hoeffler, S. (2013). Between a Rock and a Hard Place: The Failure of the Attraction Effect among Unattractive Alternatives. *Journal of Consumer Psychology*, *23*(3), 317–329. https://doi.org/10.1016/j.jcps.2012.10.008

Nickerson, R. S. (1998). Confirmation Bias: A Ubiquitous Phenomenon in Many Guises. *Review of General Psychology*, *2*(2), 175–220. https://doi.org/10.1037/1089-2680.2.2.175

NIST. (2025). *Red Team Definition*. https://csrc.nist.gov/glossary/term/Red_Team

Nodder, C. (2013). *Evil by Design: Interaction Design to Lead Us Into Temptation*. John Wiley & Sons.

Nyre-yu, M., Gutzwiller, R. S., & Caldwell, B. S. (2019). Observing Cyber Security Incident Response: Qualitative Themes from Field Research. *Proceedings of the Human Factors and Ergonomics Society*, *63*, 422–426.

O'Brien, C. E., & Katten, C. A. C. (2025, November 5). FTC's Landmark $2.5 Billion Amazon Settlement Highlights Ongoing Focus on "Dark Patterns." *The National Law Review*. https://natlawreview.com/article/ftcs-landmark-25-billion-amazon-settlement-highlights-ongoing-focus-dark-patterns

Padamwar, P. K., Kalakbandi, V. K., & Dawra, J. (2023). Deliberation Does Not Make the Attraction Effect Disappear: The Role of Induced Cognitive Reflection. *Journal of Business Research*, *154*(September 2022), 113335. https://doi.org/10.1016/j.jbusres.2022.113335

Paul, C. L., & Dykstra, J. (2017). Understanding Operator Fatigue, Frustration, and Cognitive Workload in Tactical Cybersecurity Operations. *Journal of Information Warfare*, *16*(2), 1–11.

Pettibone, J. C. (2012). Testing the Effect of Time Pressure on Asymmetric Dominance and Compromise Decoys in Choice. *Judgment and Decision Making*, *7*(4), 513–523. https://doi.org/10.1017/s1930297500002849

Pfleeger, S., & Caputo, D. (2012). Leveraging Behavioral Science to Mitigate Cyber Security Risk. *Computers & Security*, *31*(4), 597–611. http://www.sciencedirect.com/science/article/pii/S0167404811001659

Pharmer, R., Martey, R., Henery, G., Myers, E., Ray, I., & Clegg, B. (2024). Mental Firewall Breached: Leveraging Cognitive Biases for Enhanced Cybersecurity. *Human Factors in Cybersecurity*, *127*, 94–104. https://doi.org/10.54941/ahfe1004769

Pohl, R. F. (2022). Cognitive illusions: Intriguing phenomena in thinking, judgment, and memory (3rd ed.). Routledge.

Pronin, E., Gilovich, T., & Ross, L. (2002). The Bias Blind Spot: Asymmetric Perceptions of Bias in Others versus the Self. *Personality and Social Psychology Bulletin*, *28*(3), 369–381.

Roth, S., Robbert, T., & Straus, L. (2015). On the Sunk-Cost Effect in Economic Decision-Making: A Meta-Analytic Review. *Business Research*, *8*(1), 99–138. https://doi.org/10.1007/s40685-014-0014-8

Sawyer, B. D., Finomore, V. S., Funke, G. J., Mancuso, V. F., Funke, M. E., Matthews, G., & Warm, J. S. (2014). Cyber Vigilance: Effects of Signal Probability and Event Rate. *Proceedings of the Human Factors and Ergonomics Society Annual Meeting*, *58*(1), 1771–1775. https://doi.org/10.1177/1541931214581369

Simonson, I. (2014). Vices and Virtues of Misguided Replications: The Case of Asymmetric Dominance. *Journal of Marketing Research*, *51*(4), 514–519. https://doi.org/10.1509/jmr.14.0093

Singer, P., & Friedman, A. (2014). *Cybersecurity: What Everyone Needs to Know*. OUP.

Stanley, J. M., & Wedell, D. H. (2023). Impact of Choice Set Complexity on Decoy Effects. *Journal of Behavioral Decision Making*. https://doi.org/10.1002/bdm.2373

Staw, B. M. (1976). Knee-Deep in the Big Muddy: A Study of Escalating Commitment to a Chosen Course of Action. *Organizational Behavior and Human Performance*, *16*(1), 27–44. https://doi.org/10.1016/0030-5073(76)90005-2

Staw, B. M., Barsade, S. G., & Koput, K. W. (1997). Escalation at the Credit Window : A Longitudinal Study of Bank Executives ' Recognition and Write-Off of Problem Loans. *Journal of Applied Psychology*, *82*(I), 130–142.

Strauch, B. (2017). Ironies of Automation: Still Unresolved After All These Years. *IEEE Transactions on Human-Machine Systems*, *48*(5), 419–433. https://doi.org/10.1109/THMS.2017.2732506

Thaler, R. (1980). Toward a Positive Theory of Consumer Choice. *Journal of Economic Behavior and Organization*, *1*(1), 39–60. https://doi.org/10.1016/0167-2681(80)90051-7

Teovanović, P., Knežević, G., & Stankov, L. (2015). Individual differences in cognitive biases: Evidence against one-factor theory of rationality. *Intelligence*, *50*, 75–86. https://doi.org/10.1016/j.intell.2015.02.008

Trent, S., Hoffman, R. R., Leota, T., Frost, R., & Gonzalez, D. (2016). Cyberspace Operations and the People Who Perform Them. *Proceedings of the Human Factors and Ergonomics Society*, 216–217. https://doi.org/10.1177/1541931213601048

Trent, S., Hoffman, R. R., Merritt, D., & Smith, S. (2019). Modelling the Cognitive Work of Cyber Protection Teams. *The Cyber Defense Review*, *4*(1), 125–135.

Tversky, A., & Kahneman, D. (1974). Judgment under Uncertainty: Heuristics and Biases. *Science*, *185*(4157), 1124–1131. https://doi.org/10.1007/978-94-010-1834-0_8

Verizon Business. (2025). *2025 Data Breach Investigations Report.* Verizon. https://www.verizon.com/business/resources/reports/2025-dbir-data-breach-investigations-report.pdf

Wason, P. (1960). Failure to Eliminate Hypotheses in a Conceptual Task. *Quarterly Journal of Experimental Psychology*, *12*(3), 129–140. https://doi.org/10.3109/02699052.2014.965210

Wickens, C. D., Gutzwiller, R. S., & Santamaria, A. (2015). Discrete Task Switching in Overload: Two Meta-Analyses and a Model. *International Journal of Human-Computer Studies*, *79*, 79–84.

Wickens, C., McCarley, J., & Gutzwiller, R. (2022). *Applied Attention Theory* (2nd ed.). Taylor & Francis.

Wickens, C. D. (2002). Multiple Resources and Performance Prediction. *Theoretical Issues in Ergonomics Science*, *3*(2), 159–177. http://www.tandfonline.com/doi/abs/10.1080/14639220210123806

Wickens, C. D. (2008). Multiple Resources and Mental Workload. *Human Factors*, *50*(3), 449–455. https://doi.org/10.1518/001872008X288394

Wickens, C. D. (2024). The Multiple Resource Theory and Model. Some Misconceptions in Data Interpretations. *Proceedings of the Human Factors and Ergonomics Society*, 1–5. https://doi.org/10.1177/10711813241260740

Wickens, C. D., & Gutzwiller, R. S. (2017). The Status of the Strategic Task Overload Model (STOM) for Predicting Multi-Task Management. *Proceedings of the Human Factors and Ergonomics Society*, *61*, 757–761. https://doi.org/10.1177/1541931213601674

Wickens, C. D., Gutzwiller, R. S., Vieane, A., Clegg, B. A., Sebok, A., & Janes, J. (2016). Time Sharing between Robotics and Process Control: Validating a Model of Attention Switching. *Human Factors*, *58*(2), 322–343. https://doi.org/10.1177/0018720815622761

Zhong, C., Yen, J., Liu, P., Erbacher, R., Etoty, R., & Garneau, C. (2015). ARSCA: A computer tool for tracing the cognitive processes of cyber-attack analysis. *2015 IEEE International Multi-Disciplinary Conference on Cognitive Methods in Situation Awareness and Decision, CogSIMA 2015*, 165–171. https://doi.org/10.1109/COGSIMA.2015.7108193

Zhong, C., Yen, J., Liu, P., & Erbacher, R. F. (2019). Learning from Experts' Experience: Toward Automated Cyber Security Data Triage. *IEEE Systems Journal*, *13*(1), 603–614. https://doi.org/10.1109/JSYST.2018.2828832

7 Complex Healthcare Systems in the Modern Social Engineering Landscape

The Human Factors of System Survivability

Richard J. Simonson, Sarah D. Fouquet, and Crystal M. Fausett

INTRODUCTION

Social media has fundamentally reshaped how we engage with the world. From influencing purchasing decisions to shifting culture, social media has repeatedly been shown to be an agent of change. There is, however, a catch-22 regarding integrating social networking applications into our lives – the more integrated that social media is in our everyday life, the more of a cybersecurity risk it poses. A primary threat that the seamless nature of social media and real-life poses is that of social engineering or "any act that influences a person to take an action that may or may not be in their best interest" (Hadnagy, 2018). Due to the ocean of data surrounding social media, malicious actors have near immediate access to the types of sensitive information they can utilize for social engineering tactics. Under the paradigm that personal and work life are separate, this scenario may have led to a perception of low-security risk for organizations. However, that is not the case; social networks have quickly become one of the most prominent threat vectors to organizational cybersecurity (Hijji & Alam, 2021).

While this threat is domain-agnostic, the literature indicates that healthcare systems are particularly vulnerable (Nifakos et al., 2021) as evidenced by historically high rates of cyberattacks on hospital systems worldwide (Alder, 2025). Thus, we aim to address why healthcare is particularly susceptible to cyberattacks and social engineering, why social media exponentiates the frequency, intensity, and success of these attacks, and how healthcare systems must prepare themselves for the evolving cyber defense space. Throughout this chapter, we explain how healthcare employees are a prime target of malicious actors who aim to use social media as a modality for social engineering attacks, particularly targeting those

DOI: 10.1201/9781003413707-8

working on the frontline. Healthcare employees are known to work in high-stress, understaffed, and continuously evolving technological and process-based environments where errors can result in criminal charges (Lusk et al., 2022). Healthcare must grapple with the reality that social media-based platforms represent a challenging human-based gap over which organizations have little control over cyber-based protections.

This chapter discusses why this is an issue, the ramifications for healthcare and cyber defense systems, and concerns around patient safety. We detail the problems of social media and healthcare professional use, the current and evolving trends in using it as an unmoderated networking and information-sharing tool, and the damages healthcare systems have encountered from these interactions. We then describe the human factors considerations that healthcare and cyber defense organizations must plan and design around to move toward a safer and more effective healthcare environment.

TRENDS IN PERSONAL AND WORKPLACE SOCIAL MEDIA USE

Social media, initially devised as digital platforms to connect individuals remotely, have since evolved as multi-purpose tools used for various tasks, including entertainment, socialization, education, and general alleviation from boredom (Wolf, 2017). These mediums have become so ingrained into our daily lives that even new locomotion patterns have appeared: the *smartphone zombie*, someone saunters in an unengaged manner within their surroundings due to their attention toward their phone, and *Post-Flâneur* – the inconsistent and aimless wandering patterns of people who walk while using their smartphone (Argin et al., 2020). Social media usage has affected the personal lives of millions, contributing to significant challenges in areas like addiction (Hou et al., 2019), individual safety and privacy (Ružić & Jovanović, 2023), education (Selwyn & Stirling, 2016), and has even led to observed changes in brain chemistry (Korte, 2020).

Social media has undeniably altered the personal lives of everyone, regardless of one's adoption of their applications (Bullinger & Vie, 2017), and the same applies to the modern workplace. While employees can separate their personal and work lives, social media presence and use do not stop at this proverbial doorstep. For many organizations, social media is an integrated tool whose adoption has fundamentally changed the nature of work. For example, many organizations utilize social media as advertising space (St. Thomas University, 2023), as a communication and information-sharing tool, and for engagement with both their employees and customers (Queensland Government, 2022). On the other hand, organizations are encouraged to develop and enforce rules and policies for social media use during workplace activities; restrictions are placed on what employees can and cannot discuss online (Abril et al., 2012), and many industries, including healthcare, restrict phone and multimedia devices to prevent information leaks (Moses et al., 2014). While these practices effectively increase privacy and security barriers to organizations (Li et al., 2019), their effectiveness is moderated by the reality that employees' personal cyber hygiene practices follow them.

CYBERSECURITY RISKS OF PERSONAL AND WORKPLACE SOCIAL MEDIA USE

One of the most significant vulnerabilities of any cybersecurity barrier is the human element; subsequently, the most successful cyberattacks in recent history have been via the deployment of social engineering practices. The practice is implemented by tricking individuals into revealing sensitive information regarding their system to use against them or gain access (National Institute of Standards and Technology, n.d.). Social engineering is not necessarily a standardized process, but rather a series of practices for gaining access to an individual and manipulating them. Additionally, generative artificial intelligence, which further convinces users that scams are legitimate, has been identified as a highly effective exploitation method (Federal Bureau of Investigation, 2024). While no singular source comprehensively covers the vast array of methods social engineering can use, Table 7.1 details various techniques and applied examples that cybersecurity experts commonly reference (Office of Information Security, 2024; Syafitri et al., 2022).

TABLE 7.1
Common Social Engineering Practices

Method	Definition	Example Attacks	Applied Healthcare Example
Phishing	Enticing users to engage with fraudulent information to steal credentials by appearing as a reputable source.	Personal information of 200,000 people exposed when 53 public health employees clicked on a phishing link (Fry, 2024).	A fake email using an address similar to your organization's IT department requesting you to login to a URL.
Whaling	A phishing attack specifically targeting senior employees and executives.	An employee with access to all employee W-2 tax forms released said forms to a malicious actor when they received a false request from who they believed was the CEO of their organization (Gallagher, 2016).	A malicious actor targets a hospital CFO to review a contract with a new vendor via a fake URL.
Baiting	Making false promises to victims to trick users into giving information.	Researchers dispersed 297 compromised flash drives across a university campus and found that the majority of the devices were plugged into computers and their files accessed (Tischer et al., 2016).	Hospital employees receive an email from an address similar to their hospital HR with a link to a survey that promises a prize to those who take the survey.

(Continued)

TABLE 7.1 *(Continued)*
Common Social Engineering Practices

Method	Definition	Example Attacks	Applied Healthcare Example
Pretexting	Increasing the likelihood that a victim falls for a social engineering attack by providing a pretext.	A malicious actor gained access to a CEO's email and requested an account invoice change to a new bank from a partnering company (Bourke, 2025).	A cybercriminal carefully reviews social media posts of hospital CFOs. The malicious targets CFOs with carefully crafted emails based on information they gathered from posts and discussions.
Smishing	Phishing via text message.	Malicious actors sent a series of fake texts to a large number of potential victims encouraging them to download a malicious application (Cadenas, 2024).	A nurse receives a text message from a number claiming to be another nurse requesting specific and private patient details for work.
Vishing	Phishing via voice-based mediums.	Reportedly, an organization fell victim to phone calls of hackers impersonating a known employee, whose profile was found on social media, to gather sensitive login credentials to internal organization systems (Morrison, 2023).	A malicious actor calls a nurse hotline pretending to be a family member of a patient and requests private patient information.
Profile Cloning	Copying an online profile of someone who is known to the victim to trick the victim into believing them.	Malicious actors use artificial intelligence (AI)-generated audio and video on social media of known public figures to motivate potential victims in engaging in an investment scam (Butler, 2024).	A hospital employee accepts a friend request from a profile that appears to be their supervisor and begins to engage in conversation with them about work.
Shoulder Surfing	Someone who watches over the shoulder of the victim to gather personal or confidential information.	Ethical penetration testers describe their surveillance and social engineering tactics to collect data on employee credentials and patterns to gain access to secure areas (Dempsey, 2014).	A malicious actor watches over the shoulder of a provider to gather sensitive patient information.
Dumpster Diving Attack	Victims who forget or fail to properly dispose of confidential documents.	A hard drive owned by a hospital containing personal health information was left out on the front lawn after recently exchanging ownership (Jennings, 2024).	Patient information is written down on paper by providers throughout the day who accidentally leave them in public areas of a hospital.

(Continued)

TABLE 7.1 *(Continued)*
Common Social Engineering Practices

Method	Definition	Example Attacks	Applied Healthcare Example
Diversion Theft Attacks	Using courier companies known to the target organization to deliver a compromised package disguised as an expected package.	Malicious actors, posing as legitimate stakeholders, convinced a transportation company to ship three trucks worth of products into the hands of the malicious actors instead of intended retail locations (Fickenscher, 2024).	A cybercriminal masquerades as a supply chain manager for a hospital and convinces the courier to transport crucial supplies to the attackers' desired location.
Tailgating	Following a credentialed user to gain physical or virtual access to protected systems.	Storied example of a penetration tester gaining access into a facility by posing as a delivery driver (Datacenters.com, 2023).	Someone walks near a distracted healthcare employee and follows them through security doors to gain access to confidential areas.
File Masquerade	Taking advantage of the fact that typical users will not be familiar with every file and application on their computer to upload malicious files.	Using malicious code, computer users were tricked into a fake file with a hidden extension that non-tech savvy users would be unaware to check (Goodin, 2024).	A provider downloads a series of PDF scholarly articles for work but is unaware that one file had a hidden executable extension.
Reverse Social Engineering	Creating challenges wherein the user will accept help from a seemingly reputable source without questioning it.	A victim sends nearly 500 million pesos in gift cards to a malicious actor posing as the victim's bank (Lugto, 2024).	A hospital employee receives an email from an address that looks like their IT department offering them a link to a new program that will speed up their computer.

SOCIAL MEDIA USE AND WORKPLACE RISKS

As discussed, social media does not have a proverbial front door in which the employee's personal and work lives are separate. Instead, employee social media profiles represent a window, and often a bibliographic camera, of a person's life that can reveal the sensitive personally identifying information and personal beliefs of the account holder and anyone with whom the account holder uploads content. In other words, it is a source that provides anyone with the resources they would need to target and manipulate for malicious reasons with something as simple as a friend request (Bilge et al., 2009).

Increased rates and depth of interaction with social media can influence the cyber hygiene of its users. Literature on social media in personal security matters shows a clear relationship; increased engagement with social media is correlated with increased security vulnerability (Albladi & Weir, 2020). While social media sites provide mechanisms to increase the privacy and security of their users, such as informing users of risks, preventing users from uploading/sharing sensitive information, and providing recommendations for security measures (Li et al., 2019; Venkatanathan et al., 2014), they depend on the user employing them (Rathore et al., 2017) and the user's own choices to share such information (Albladi & Weir, 2020). Additionally, the multimedia aspect of social media platforms encourages users to upload text descriptions, audio, and video of themselves, their surroundings, and others that may inadvertently reveal sensitive and personally identifiable information (Hamid et al., 2022). These are challenges that the modern organization's cybersecurity policies and practices, particularly of smaller social enterprises, are not prepared for (White et al., 2022). They are also challenges that are only increasing in both frequency and intensity, as approximately 22% of all social engineering attacks originate from social media apps (Hijji & Alam, 2021), with workplace social media use quickly rising to the primary network entry mechanisms for malicious actors (Aldawood & Skinner, 2018). Underdeveloped and resource-limited cyber defense mechanisms of healthcare organizations have the potential to suffer greater likelihood and intensity of these techniques.

HEALTHCARE: TARGET RICH ENVIRONMENTS

Worldwide, healthcare systems suffer from the underdevelopment and underfunding of their cyber defense capabilities (He et al., 2021) and their integration of cyber hygiene training and practices with employees (Argyridou et al., 2023). Hospitals and other healthcare systems have been the subject of numerous cyberattacks and data breaches annually. A global healthcare survey found that over 33% of all healthcare systems have suffered at least one ransomware attack (Mishra, 2024). These attacks are frequently increasing, with healthcare systems globally experiencing 1463 attacks per week in 2022 and 1500 per week in 2023 (Petrosyan, 2024). This trend has continued, with some sources reporting 2018 weekly attacks in 2024, an estimated 32% increase from 2023 (Ribeiro, 2024).

The motives behind this attack are clear: gaining access to patient health data and holding healthcare systems ransom is lucrative; patient health data can be sold illegally for up to $1000 per record, depending on whose records they are (Owens, 2020). Healthcare systems tend to make easy targets for malicious actors due to the time-sensitive nature of the information. Subsequently, these records usually have lesser internal protective measures for quicker accessibility (Bris & Asri, 2016). Another factor affecting the vulnerability of healthcare systems to social engineering-based cyberattacks is the particular susceptibility of its employees. While staff in these domains do not significantly differ from their counterparts in terms of training and competence, they are subjugated to multiple critical factors known to decrease cognitions behind the detection and recognition of social engineering attacks, including (1) high-stress environments, (2) high levels of workload, (3) susceptibility to increased fatigue, (4) challenges in organizational culture, and (5) high levels of burnout.

Stress

Healthcare employees have consistently reported exceptionally high occupational stress levels (American Hospital Association, n.d.; Bayram Deger, 2024; CDC, 2023, 2024; Rink et al., 2023), contributed by the subjugation to organizational culture (Khatri et al., 2009), death and suffering from occupational hazards (Rigas et al., 2024), and long work hours (Caruso, 2014). Unabated occupational stress levels contribute to numerous adverse outcomes, including higher rates of burnout and suicide (Abbasi, 2022; Joseph, 2022). High-stress levels are also known to decrease both work-related performances (Staal, 2004) and safety-related behaviors (Wang et al., 2018). Thus, it should be no surprise that employees who rate their perceived occupational stress as high fall victim to social engineering tactics more frequently than those who rate their perceived stress as low (Bell et al., 2019; Nifakos et al., 2021).

Workload

Healthcare systems worldwide face a staffing shortage at astonishing levels with unachievable task loads. For example, studies on primary care find that physicians would need to spend, on average, 26.7 hours per day to provide optimum care to the number of patients they see yearly (Porter et al., 2023); hospital-based physicians experience high workload levels to the point where an estimated 83% are at or above full workload capacity (The Physicians Foundation, 2024). The nursing shortage has been in full force for nearly 90 years (University of Pennsylvania, n.d.) and has recently been exacerbated by the 2019 SARS-COVID pandemic (Lucchini et al., 2020). Such factors contribute to healthcare staff's increasingly high workload levels, significantly affecting their ability to recognize and detect malicious information and social engineering practices (Jalali et al., 2020).

Fatigue

Fatigue is a psychological state associated with many occupation risks and has been the direct source of severe and fatal occupational accidents across many domains (Williamson et al., 2011) and described a complex psychological and physiological mechanism whose characteristics negatively affect various performance-based tasks (Behrens et al., 2023). Fatigue affects human operators by reducing their abilities to think, move, and be aware of themselves and their surroundings (Abd-Elfattah et al., 2015). This is caused primarily by the body not receiving enough quality sleep (Mayo Clinic, n.d.). Fatigue is a phenomenon that runs rampant in the healthcare sector due to long work hours and understaffing (Owens, 2007). It is so rampant that it contributes to approximately 82% of medication-related errors (Bell et al., 2023). However, fatigue is a complex phenomenon whose multi-faceted nature affects healthcare employees and their abilities to practice cyber hygiene in various ways.

The above narrative primarily focuses on sleep-related fatigue, which, as established, affects conditions like decision-making and one's ability to detect and recognize abnormal information. However, two other forms of fatigue can also affect an employee's ability to prevent or mitigate a social engineering attack. In healthcare, frontline providers often experience a phenomenon called compassion fatigue, which, in the context of this chapter, describes the feeling of not being able to provide the necessary care to a patient (Ledoux, 2015). This type of fatigue is associated with decreased cyber awareness and cyber hygiene because providers disregard certain risks of their decisions and actions to prioritize patient care (Klindienst et al., 2022). Another form of fatigue, security fatigue, describes a psychological state in which an employee is resistant to security policies and compliance due to interruptions in their ability to complete their work (Stanton et al., 2016). In healthcare, this fatigue construct ties together the staff's ability to (1) care about cyber threats over the state of their patients and (2) recognize, react, and act appropriately to cyber threats such as social engineering attacks.

Organizational Culture

Organizational culture has heterogeneous defining characteristics, with one review of the extant literature finding 54 varying definitions of the construct (Verbeke et al., 1998). Within the context of this chapter, we define organizational culture according to Schein (1984) who states:

> Organizational culture is the pattern of basic assumptions that a given group has invented, discovered, or developed in learning to cope with its problems of external adaptation and internal integration and that has worked well enough to be considered valid and, therefore, to be taught to new members as the correct way to perceive, think, and feel about those problems.
>
> **p. 3**

Organizational culture plays a significant role in an organization's capabilities and vulnerabilities regarding cyberattacks via network intrusions and social engineering practices (Handri et al., 2024). For one, healthcare systems employ highly motivated and engaged staff to serve and care for their patients, often putting their mental and physical well-being on hold (Kalmoe et al., 2019). However, the cultivation of these cultures can affect how staff detect, react to, or care for their organization's security over patient care. For example, Nifakos et al. (2021) and Skyvell Nilsson et al. (2018) found healthcare providers were more likely to disregard security policies and concerns if they significantly impeded patient care.

Culture can also impede cybersecurity defensive capabilities through policies and practices focusing on accountability and continuous improvement. Healthcare is a high-risk domain whose science of practice is constantly evolving. Between growing health-related misinformation on social media (Gaysynsky et al., 2024), extraordinarily high rates of global healthcare census (Humphreys et al., 2022), a severe shortage of healthcare providers (Hwang & Woodrum, 2020), and supply chain challenges that introduce variability of performance and care (Kanokphanvanich et al., 2022), providers are incessantly bombarded with new processes, protocols, and policies. When things go wrong in these systems, there exists a tendency to blame the human rather than identify the complex factors that led to the failure and encourage staff to hide, rather than report, errors and mistakes (Catino, 2009), contributing to decreased security mindsets (Hüsch et al., 2024).

CASES OF SOCIAL ENGINEERING ATTACKS ON HEALTHCARE VIA SOCIAL MEDIA

The accessibility of multimedia-based social media to employees in the workplace has provided a data-rich environment for malicious actors to gain information about healthcare employees' personal lives, their networks of friends and colleagues, their patients and other stakeholders, and even access to their organizational structure. Unsurprisingly, multiple cases of social engineering attacks on health systems directly using or gathering information from employee and employee-adjacent social media accounts have been purposefully studied and publicly recorded.

Phishing Emails and Public Employee Profiles

A study by Priestman et al. (2019) assessed their hospital staff's resilience to mitigate a mass phishing attack. The authors contracted a third-party cybersecurity firm to conduct a cybersecurity audit and a simulated phishing attack. Fortunately, the authors report that no internal information was accessed due to security design and employee vigilance. However, the audit process identified 486 employee credentials by scraping publicly available data online. Sources included social media, where some employees were found to have both accepted friend requests from fake profiles and engaged in direct messaging with the fake profile. Concerningly, employee profiles on dating apps included personal photos with badge credentials and uniforms that were visible.

Boston Children's Hospital

In late 2013, the now-publicized case of 16-year-old Justina Pelletier began to circulate the news describing her experience of a custody dispute between the State of Massachusetts and her family (Gardner et al., 2016; Wagner, 2014). Initiated by a disagreement in diagnosis and treatment (Gardner et al., 2016; Wagner, 2014), Justina spent more than a year in state custody receiving care at Boston Children's Hospital (BCH). During her stay, local and international bad actors began initiatives to show support for either side through various mechanisms, including malicious targeting of the BCH network. These attacks became crowdsourced as the staff and providers directly involved in Justina's cases were posted publicly online, allowing anyone with intent to further identify others in the hospital or engage with the staff themselves directly.

Impersonating Government Officials

In 2022, financial extortion and personally identifiable information (PII) theft scams plagued social media apps targeting care providers (Federal Bureau of Investigation Baltimore, 2022). The fraud consisted of a network of malicious actors that utilized publicly available data through various modalities (e.g., social media and organization websites) to curate information on intended targets while depicting themselves as legitimate government officials and law enforcement officers to demand money or access to patient PII records.

DEVELOPING A SURVIVABLE CYBER SYSTEM IN THE AGE OF SOCIAL MEDIA DEPENDENCY

An organization's ability to fend off cyberattacks and maintain its operational state is rooted in the survivability of its systems. Explicitly defined as a key performance parameter of organizational performance via preventative measures, mitigating acts, and attack recovery, system survivability is a crucial mechanism to any domain, as indicated by the framework's wide extension in multiple high-reliability domains (Defense Acquisition University, n.d.-c). However, to truly encapsulate the concept of system survivability and its place in reducing the effectiveness of social engineering practices in the data through social media, we first must understand what systems are and how they are defined.

SYSTEMS SCIENCE AND ASSESSMENTS

Systems are generically defined as "groups or combinations of interrelated, interdependent, or interacting elements forming collective entities" (Arnold & Wade, 2015, p. 675). We state *generically* because any singular definition cannot epitomize the complexity of systems. Instead, we suggest that systems are characterized by any manifest or latent interaction that influences processes. In extending this to an outcome-based view, systems can be described by a network of interacting inputs that affect processes and generate outcomes. The healthcare sector often references the Systems Engineering Initiative for Patient Safety (SEIPS) model to visualize the complexity of systems (Carayon et al., 2020). The SEIPS model helpfully breaks down these three categories, inputs, processes, and outputs, into digestible categories that are easy to understand and convey. Understanding systems is essential to implementing system survivability not just because the word system is in the name but because the development of preventative, mitigating, and recovery strategies depends on one's understanding of how humans form and act on decisions.

The work system components of the system characterize the agents that interact to create processes, SEIPS 2.0 breaks down the work system into six categories. Tools and technologies are the characteristics of the things that are used to create work, including hardware, software, and physical items. Characteristics of tools and technologies might include differences between software that requires a physical badge to login and software that requires a password; the interface describes differing interactions with differing experiences of usability and accessibility. Organizations are the characteristics that describe the structure and policy of the system. These characteristics might include the cybersecurity training policies and programs for employees. Tasks describe the individuals' steps required to accomplish a greater goal. For example, employees don't simply avoid phishing emails, rather, they have to open, read, recognize, and report the email to successfully defeat a potential phishing attack. The internal environment describes the characteristics of the physical environment; these might include ambient temperature (e.g., too hot and too cold) or physical barriers that prevent unintended access. The external environments are things that are greater to the immediate system, but still affect the system to a large extent. These might include cultural, policy, and law differences between healthcare systems and borders. Finally, person(s) describes the characteristics of both individuals and teams. These might describe cognitive (e.g., trust, psychological safety, stress, fatigue, knowledge, skills, and attitudes) and physical states (e.g., height and strength).

Processes describe the cognitive, physical, and social or behavioral interactions with the work system characteristics to produce work. An easy way to think about processes is, from our tools and technologies example above, from a work system perspective a physical badge may be easier to use than a password; processes instead are the decisions and physical actions the employee must take to use the badge reader or password. The badge would require the employee to remember to bring their badge and accurately touch the badge to the reader. Whereas password-based entry would require the employee to remember their password, type their password into their keyboard. Finally, outcomes describe the state of the system as a result of work, which cyclically defines the characteristics of the work system.

System Assessments

Taking a holistic view of the cybersecurity threat landscape is essential because it helps organizations understand and consider all external and internal factors that might increase their vulnerability to and resilience against social engineering attacks. To achieve this, a comprehensive systems needs assessment should be the first step for any organization or cybersecurity professional. This assessment offers guidance on risk factors that need to be addressed and identifies existing protocols, policies, and integrated technologies that may require further attention.

While there are numerous methods available for conducting such assessments, as outlined by Leszczyna (2021), who categorized 32 methods into compliance, vulnerability, penetration testing, and simulation testing, a holistic approach that includes all components of the system is a fundamental requirement. Thus, independent of any method, we highly recommend the integration of the SEIPS (Holden et al., 2013), or similar, work system model to help drive the questions and considerations for identifying needs. A helpful resource in this endeavor includes the National Health Service (NHS) work system explorer (2022). Once a comprehensive assessment cycle is complete, which will be an iterative and continuously repeated process, the identified risk factors should be resolved via their integration into the system survivability framework.

System Survivability: Prevention, Mitigation, and Recovery

A system's survivability is predicated on an organization's ability to prevent events, mitigate attacks when they occur, and recover from their effects to return to operational standards quickly. In other words, system survivability is a framework that ingrains an organization's ability to get hit and continue moving, which predictably is a crucial characteristic of healthcare systems whose patient lives depend on them. However, the integration of the system survivability framework necessitates a comprehensive approach that accounts for both the technical network components and the human components of the system. As this chapter focuses on the human factors of system performance, we primarily focus on system survivability frameworks and mechanisms for healthcare system staff. Resources on system survivability of networks and other technical aspects of cybersecurity are numerous (Defense Acquisition University, n.d.-a). They should be referenced in combination with the concepts covered in this chapter.

Susceptibility and Prevention

It's said that "an ounce of prevention is worth a pound of cure" (Franklin, 1735). This phrase applies to a system's survivability as much as it did in Benjamin Franklin's advisory on fire prevention. Within the system survivability framework, prevention is a key factor that aims to reduce an organization's susceptibility to some attack (Defense Acquisition University, n.d.-c). Integrating prevention into cybersecurity systems is a crucial aspect of initial system design and cannot be determined by testing a system's ability to prevent planned or preconceived attacks (Barnett, 2022). Due to cyberattacks' dynamic and constantly evolving nature, the susceptibility of attacks must be analyzed via various systems assessment methods by domain-specific, systems engineering, and human factors experts throughout the conception and design of said system.

From a human factors perspective, systems assessments aim to identify the factors contributing to human performance and account for these factors within the system's design. These steps are elucidated by the Defense Acquisition University (n.d.-a) product 10-23-1:

1. Conduct a systems assessment.
2. Evaluate the proposed (or current) system design and assess its survivability with historical, current, and projected operational environments.
3. Conduct a quantifiable risk assessment of identified system factors.
4. Calculate the cost/benefit of preventative measures over operational cost and expected risk.
5. Provide systems recommendations regarding risk and cost/benefit ratios.
6. Document identified risks.
7. Incorporate risk profiles.

End users, a central system category, represent one of the most susceptible components within the security framework. Thus, we recommend focusing on methodologies that assess system design and its vulnerabilities by how they characteristically influence users to act and react. These assessments primarily focus on humans' cognitive limitations within their systems and goals.

Multiple prevention models, summarized by Aldawood and Skinner (2018) and Syafitri et al. (2022), have been developed in response to these complex challenges. However, prevention relies on barriers and processes already developed to help stop the error before it occurs. Unfortunately, such a task is impossible as organizations cannot be aware of the unobservable errors or the errors that occur without being reported, as shown in recent medication prescription literature (Li et al., 2024). Subsequently, we must also ensure that our systems are designed to help users identify an error that occurs (Morey et al., 2023) and provide the resources necessary such that the user can ensure the error does not affect the patient, team member, or organization. Thus, we must focus on understanding that all systems are vulnerable to attack and will eventually fall victim to malicious actors. This framework describes vulnerability as the system's characteristics (refer to the SEIPS work system, Holden et al., 2013). Systems whose designs cannot or do not account for specific threat vectors will fail to prevent attacks from breaking through preventative barriers.

Mitigation

It's important to note that attacks that reach the mitigation stage have not yet affected the system. Instead, they represent a state where the system has failed to prevent an attack and may experience performance degradation (Defense Acquisition University, n.d.-b). At this stage of system performance, the system must be designed to bring users' attention to these threats and assist the user in stopping the threat before damage can occur. As discussed in prior sections, healthcare systems' security weaknesses are built in by design, as staff must have quick and readily available access to sensitive patient and organizational information. Fortunately, advances in cybersecurity techniques and strategies have significantly decreased the vulnerability of these systems to attacks (Priestman et al., 2019). However, attacks that originate outside of organization-controlled networks still offer vulnerabilities that malicious actors often target. The availability of personal and personal-adjacent information by staff members on social media is a significant threat to individual and organizational security (Bilge et al., 2009).

Mitigating social engineering attacks can be more complex than preventing them, as organizations and individuals must first gain awareness that an attack has been initiated. However, multiple techniques have been established, primarily via technologies, to ensure an attack cannot reach internal systems or via network monitoring (Zulkurnain et al., 2015). Human-centric mitigation models are available that encourage the integration of all employees into the social engineering mitigation process. For example, the user-reflective mitigation model employs an organizational strategy that trains employees to reflect on their interactions to identify abnormalities and determine if a social engineering attack is possible (Airehrour et al., 2018), a crucial stage as it prompts many subsequent mitigating strategies, such as a cybersecurity response team (Samad, 2024). It is entirely dependent on the recognition of said attack. Due to the highly dynamic nature of social engineering practices and the complexity of modern networking systems, neither the human nor technological approach to social engineering mitigation is independently practical. Instead, they must be integrated as a cohesive approach, continuously stress-tested, and updated according to the threat space (Osuagwu et al., 2015).

Recoverability

The system survivability tenet of recovery describes systems that can continue operating while aiming to return to normality under an actual threat to the system (Defense Acquisition University, n.d.-b). It might sound as if recovery is a stage only active when a system is under threat, which, according to Linger et al. (2002), is a misnomer of recoverability who argues that no system exists solely in a binary state of safe and compromised. Instead, like prevention and mitigation, this tenet is a characteristic ingrained into the system. Any compromise allows the system to continue operating, albeit at a reduced rate. Consequences of improperly designed recoverability can lead to significant disruptions in care (Hannah et al., 2024), closures of hospitals (Collier, 2023), and even patient fatalities (Miller, 2022). Fortunately, healthcare staff are no strangers to error recovery. For example, literature assessing the impact of frontline providers on errors finds that nurse professionals recover up to two errors per shift (Gaffney et al., 2016). The reality of the frontline employees during cyber-based outages is that they are the final line between complete system failure and continued operations. For example, Princeton Community Hospital, West Virginia quickly adapted to a cybersecurity attack by integrating its cyber response system by relying on the training and expertise of its staff to use non-computer methods for patient care and information transfer (The Joint Commission, 2023). Recovery strategies depend on organizational capabilities and the mechanism and type of attack. Generally, three immediate steps are required: (1) prevent the spread of the attack by isolating the affected systems, (2) communicate with key stakeholders to avoid confusion and prevent the attack from unexpectedly spreading, and (3), if possible, restore affected systems and utilize fallback methods to continue system-critical tasks (Samad, 2024).

CONCLUSION

The integration of social media into the workplace has introduced both opportunities and significant risks that healthcare organizations must navigate carefully. The pervasive nature of social media profiles acts as a window into employees' personal lives, offering malicious actors a wealth of information to exploit through social engineering attacks. From phishing attempts to impersonation scams, these attacks target healthcare workers and their access to sensitive patient data, ultimately compromising organizational security and patient trust. Moreover, the vulnerabilities exacerbated by high levels of occupational stress, overwhelming workloads, fatigue, and organizational culture create an environment where employees are more susceptible to falling victim to such attacks.

The challenge for healthcare systems lies in their underdeveloped cyber defense capabilities, limited resources, and the rising sophistication of social engineering tactics. As malicious actors increasingly exploit the weaknesses of healthcare environments, the consequences extend beyond data breaches—impacting patient care, operational stability, and the organization's reputation. While healthcare organizations have implemented various cybersecurity measures, they often fall short in addressing the unique vulnerabilities introduced by social media. The lack of awareness, insufficient training, and failure to cultivate a security-first culture further compound these challenges.

To create a survivable cybersecurity system, healthcare organizations must adopt a multi-faceted approach that combines technical defenses with a thorough understanding of human behavior. This involves enhancing employee awareness and providing robust cybersecurity training that not only targets technical skills but also addresses the psychological and behavioral aspects of security. By fostering a culture of vigilance, promoting best practices for cyber hygiene, and encouraging proactive reporting of suspicious activities, healthcare workers can better detect and respond to emerging threats.

Moreover, preventing social engineering attacks requires a comprehensive system of preemptive measures, mitigation strategies, and recovery frameworks. The healthcare sector must adopt a systems-thinking approach—viewing cybersecurity as an interconnected web of tools, processes, organizational structures, and individual behaviors that collectively contribute to system survivability. This means understanding the interaction between people, technology, and the work environment, as well as continuously refining strategies to account for new vulnerabilities and threats.

Ultimately, the key to safeguarding healthcare systems lies in building resilience—balancing technological defenses with the critical role played by the human factor. Healthcare organizations must move beyond reactive cybersecurity measures and embrace a proactive, holistic approach to protecting patient information and ensuring the continued integrity of healthcare services. Only through a concerted effort to address these vulnerabilities can healthcare systems create a resilient cybersecurity posture capable of withstanding the evolving landscape of social engineering attacks.

REFERENCES

Abbasi, J. (2022). Pushed to their limits, 1 in 5 physicians intends to leave practice. *JAMA*, *327*(15), 1435–1437. https://doi.org/10.1001/jama.2022.5074

Abd-Elfattah, H. M., Abdelazeim, F. H., & Elshennawy, S. (2015). Physical and cognitive consequences of fatigue: A review. *Journal of Advanced Research*, *6*(3), 351–358. https://doi.org/10.1016/j.jare.2015.01.011

Abril, P. S., Levin, A., & Del Riego, A. (2012). Blurred boundaries: Social media privacy and the twenty-first-century employee. *American Business Law Journal*, *49*, 63.

Airehrour, D., Vasudevan Nair, N., & Madanian, S. (2018). Social engineering attacks and countermeasures in the New Zealand banking system: Advancing a user-reflective mitigation model. *Information*, *9*(5), Article 5. https://doi.org/10.3390/info9050110

Albladi, S. M., & Weir, G. R. S. (2020). Predicting individuals' vulnerability to social engineering in social networks. *Cybersecurity*, *3*(1), 7. https://doi.org/10.1186/s42400-020-00047-5

Aldawood, H., & Skinner, G. (2018). Educating and raising awareness on cyber security social engineering: A literature review. *2018 IEEE International Conference on Teaching, Assessment, and Learning for Engineering (TALE)*, 62–68. https://doi.org/10.1109/TALE.2018.8615162

Alder, S. (2025, January 7). The biggest healthcare data breaches of 2024. *The HIPAA Journal*. https://www.hipaajournal.com/biggest-healthcare-data-breaches-2024/

American Hospital Association. (n.d.). *Supporting behavioral health | Strengthening the health care workforce | AHA*. American Hospital Association. Retrieved February 4, 2025, from https://www.aha.org/workforce-strategies/supporting-behavioral-health

Argin, G., Pak, B., & Turkoglu, H. (2020). Between post-flâneur and smartphone zombie: Smartphone users' altering visual attention and walking behavior in public space. *ISPRS International Journal of Geo-Information*, *9*(12), Article 12. https://doi.org/10.3390/ijgi9120700

Argyridou, E., Nifakos, S., Laoudias, C., Panda, S., Panaousis, E., Chandramouli, K., Navarro-Llobet, D., Zamorano, J. M., Papachristou, P., & Bonacina, S. (2023). Cyber hygiene methodology for raising cybersecurity and data privacy awareness in health care organizations: Concept study. *Journal of Medical Internet Research*, *25*(1), e41294. https://doi.org/10.2196/41294

Arnold, R. D., & Wade, J. P. (2015). A definition of systems thinking: A systems approach. *Procedia Computer Science*, *44*, 669–678. https://doi.org/10.1016/j.procs.2015.03.050

Barnett, T. (2022, March 17). *Applying the tenets of aircraft combat survivability to achieve cyber resilience* [PowerPoint]. https://www.ndia.org/-/media/sites/ndia/divisions/combat-survivability/2022-webinar-slides/applying-tenets-of-aircraft-combat-survivability-cyber-resilience-v11--barnett.pdf

Bayram Deger, V. (2024). Editorial: Anxiety, burnout, and stress among healthcare professionals. *Frontiers in Psychology*, *14*. https://doi.org/10.3389/fpsyg.2023.1348250

Behrens, M., Gube, M., Chaabene, H., Prieske, O., Zenon, A., Broscheid, K.-C., Schega, L., Husmann, F., & Weippert, M. (2023). Fatigue and human performance: An updated framework. *Sports Medicine*, *53*(1), 7–31. https://doi.org/10.1007/s40279-022-01748-2

Bell, A. J. C., Rogers, M. B., & Pearce, J. M. (2019). The insider threat: Behavioral indicators and factors influencing likelihood of intervention. *International Journal of Critical Infrastructure Protection*, *24*, 166–176. https://doi.org/10.1016/j.ijcip.2018.12.001

Bell, T., Sprajcer, M., Flenady, T., & Sahay, A. (2023). Fatigue in nurses and medication administration errors: A scoping review. *Journal of Clinical Nursing*, *32*(17–18), 5445–5460. https://doi.org/10.1111/jocn.16620

Bilge, L., Strufe, T., Balzarotti, D., & Kirda, E. (2009). All your contacts are belong to us: Automated identity theft attacks on social networks. *Proceedings of the 18th International Conference on World Wide Web*, 551–560. https://doi.org/10.1145/1526709.1526784

Birthriya, S. K., Ahlawat, P., & Jain, A. K. (2025). A comprehensive survey of social engineering attacks: taxonomy of attacks, prevention, and mitigation strategies. *Journal of Applied Security Research*, *20*(2), 244–292.

Bourke, K. (2025, January 15). Company scammed out of $192k via fake invoice ordered to pay bill. *ABC News*. https://www.abc.net.au/news/2025-01-16/court-orders-inoteq-to-pay-190k-after-fraudulent-invoice/104783454

Bris, A. L., & Asri, W. E. (2016). *State of cybersecurity & cyber threats in healthcare organizations*. ESSEC Business School. https://f.hubspotusercontent00.net/hubfs/8011857/Admere_August2020/Images/risks-and-threats-healthcare-strategic-report.pdf

Bullinger, C., & Vie, S. (2017). After a decade of social media: Abstainers and ex-users. In D. M. Walls & S. Vie (Eds.), *Social writing/social media: Publics, presentations, and pedagogies* (pp. 69–88). The WAC Clearinghouse; University Press of Colorado. https://doi.org/10.37514/PER-B.2017.0063.2.04

Butler, J. (2024, June 7). Deepfakes of Australian politicians including Penny Wong and Katy Gallagher used in investment scams. *The Guardian*. https://www.theguardian.com/australia-news/article/2024/jun/08/deepfakes-of-australian-politicians-including-penny-wong-and-katy-gallagher-used-in-investment-scams

Cadenas, C. (2024, August 3). *Cybersecurity firm warns Android users to watch out for money-draining malware*. TechRadar. https://www.techradar.com/computing/software/cybersecurity-firm-warns-android-users-to-watch-out-for-money-draining-malware

Carayon, P., Wooldridge, A., Hoonakker, P., Hundt, A. S., & Kelly, M. M. (2020). SEIPS 3.0: Human-centered design of the patient journey for patient safety. *Applied Ergonomics*, *84*, 103033. https://doi.org/10.1016/j.apergo.2019.103033

Caruso, C. C. (2014). Negative impacts of shiftwork and long work hours. *Rehabilitation Nursing: The Official Journal of the Association of Rehabilitation Nurses*, *39*(1), 16–25. https://doi.org/10.1002/rnj.107

Catino, M. (2009). Blame culture and defensive medicine. *Cognition, Technology & Work*, *11*(4), 245–253. https://doi.org/10.1007/s10111-009-0130-y

CDC. (2023, October 24). *Health worker mental health crisis*. Centers for Disease Control and Prevention. https://www.cdc.gov/vitalsigns/health-worker-mental-health/index.html

CDC. (2024, August 8). *Risk factors for stress and burnout*. Healthcare Workers. https://www.cdc.gov/niosh/healthcare/risk-factors/stress-burnout.html

Collier, K. (2023, June 12). *An Illinois hospital links closure to ransomware attack*. NBC News. https://www.nbcnews.com/tech/security/illinois-hospital-links-closure-ransomware-attack-rcna85983

Datacenters.com. (2023, October 27). *Social engineering impacts physical security at data centers*. Datacenters.Com. https://www.datacenters.com/news/social-engineering-impacts-physical-security-at-data-centers

Defense Acquisition University. (n.d.-a). *SE Brainbook Home*. Systems Engineering Brainbook. Retrieved January 7, 2025, from https://content1.dau.edu/DAUMIG_se-brainbook_189/content/index.html

Defense Acquisition University. (n.d.-b). *Survivability*. Defense Acquisition University. Retrieved February 4, 2025, from https://content1.dau.edu/DAUMIG_se-brainbook_189/content/Design%20Considerations/Survivability.html

Defense Acquisition University. (n.d.-c). *System survivability key performance parameter (KPP) (mandatory KPP)* | www.dau.edu. Retrieved February 4, 2025, from https://www.dau.edu/glossary/system-survivability-ss-key-performance-parameter

Dempsey, M. (2014, October 10). *Meet the team breaking into top-secret HQ's*. BBC. https://www.bbc.com/news/articles/c8el64yyppro

Federal Bureau of Investigation. (2024, December 3). *Internet Crime Complaint Center (IC3) | Criminals use generative artificial intelligence to facilitate financial fraud*. Public Service Announcement. https://www.ic3.gov/PSA/2024/PSA241203#fn3

Federal Bureau of Investigation Baltimore. (2022). *FBI warns individuals employed in the health care industry of the ongoing scam involving the impersonation of law enforcement and government officials* [Press Release]. Federal Bureau of Investigation. https://www.fbi.gov/contact-us/field-offices/baltimore/news/press-releases/fbi-warns-individuals-employed-in-the-healthcare-industry-of-the-ongoing-scam-involving-the-impersonation-of-law-enforcement-and-government-officials

Fickenscher, L. (2024, December 16). Exclusive | Holiday heist diverts $1M in toys on 3 different trucks in new type of scam. *New York Post*. https://nypost.com/2024/12/16/business/holiday-heist-diverts-1m-in-toys-on-3-different-trucks-in-new-type-of-scam/

Franklin, B. (1735, February 4). *Founders online: On protection of towns from fire, 4 February 1735*. National Archives; University of Virginia Press. http://founders.archives.gov/documents/Franklin/01-02-02-0002

Fry, H. (2024, June 15). Phishing attack hits L.A. County public health agency, jeopardizing 200,000-plus residents' personal info. *Los Angeles Times*. https://www.latimes.com/california/story/2024-06-14/la-county-public-health-data-breach-possibly-affects-200-000-are-you-one-of-them

Gaffney, T. A., Hatcher, B. J., & Milligan, R. (2016). Nurses' role in medical error recovery: An integrative review. *Journal of Clinical Nursing*, *25*(7–8), 906–917. https://doi.org/10.1111/jocn.13126

Gallagher, S. (2016, March 7). *Seagate employees' W-2 forms exposed in another payroll phish*. Ars Technica. https://arstechnica.com/information-technology/2016/03/seagate-employees-w-2-forms-exposed-in-another-payroll-phish/

Gardner, K., Ruest, S., & Cummings, B. (2016). Diagnostic uncertainty and ethical dilemmas in medically complex pediatric patients and psychiatric boarders. *Hospital Pediatrics*, *6*(11), 689–692. https://doi.org/10.1542/hpeds.2016-0032

Gaysynsky, A., Everson, N. S., Heley, K., & Chou, W.-Y. S. (2024). Perceptions of health misinformation on social media: Cross-sectional survey study. *JMIR Infodemiology*, *4*(1), e51127. https://doi.org/10.2196/51127

Goodin, D. (2024, July 10). *Threat actors exploited Windows 0-day for more than a year before Microsoft fixed it*. Ars Technica. https://arstechnica.com/security/2024/07/threat-actors-exploited-windows-0-day-for-more-than-a-year-before-microsoft-fixed-it/

Hadnagy, C. (2018). *Social engineering: The science of human hacking* (2nd ed.). Wiley. https://www.wiley.com/en-us/Social+Engineering%3A+The+Science+of+Human+Hacking%2C+2nd+Edition-p-9781119433750

Hamid, A., Alam, M., Sheherin, H., & Pathan, A.-S. K. (2022). Cyber security concerns in social networking service. *International Journal of Communication Networks and Information Security (IJCNIS)*, *12*(2). https://doi.org/10.17762/ijcnis.v12i2.4634

Handri, E. Y., Sensuse, D. I., & Lusa, S. (2024). Examining cybersecurity culture: Trends and success factors. *Journal of Internet Services and Information Security*, *14*(3), 330–352. https://doi.org/10.58346/JISIS.2024.I3.020

Hannah, J., Murphy, T., & Foody, K. (2024, May 10). *A cyberattack forces a big US health system to divert ambulances and take records offline.* AP News. https://apnews.com/article/cyberattack-hospital-system-ambulances-diverted-ascension-728ab2a0e5afaf7c344e46a5ce5ca42c

He, Y., Aliyu, A., Evans, M., & Luo, C. (2021). Health care cybersecurity challenges and solutions under the climate of COVID-19: Scoping review. *Journal of Medical Internet Research*, *23*(4), e21747. https://doi.org/10.2196/21747

Hijji, M., & Alam, G. (2021). A multivocal literature review on growing social engineering based cyber-attacks/threats during the COVID-19 pandemic: Challenges and prospective solutions. *IEEE Access*, *9*, 7152–7169. IEEE Access. https://doi.org/10.1109/ACCESS.2020.3048839

Holden, R. J., Carayon, P., Gurses, A. P., Hoonakker, P., Hundt, A. S., Ozok, A. A., & Rivera-Rodriguez, A. J. (2013). SEIPS 2.0: A human factors framework for studying and improving the work of healthcare professionals and patients. *Ergonomics*, *56*(11), 1669–1686. https://doi.org/10.1080/00140139.2013.838643

Hou, Y., Xiong, D., Jiang, T., Song, L., & Wang, Q. (2019). Social media addiction: Its impact, mediation, and intervention | Cyberpsychology. *Journal of Psychosocial Research on Cyberspace Cyberpsychology*, *13*(1). https://cyberpsychology.eu/article/view/11562

Humphreys, P., Spratt, B., Tariverdi, M., Burdett, R. L., Cook, D., Yarlagadda, P. K. D. V., & Corry, P. (2022). An overview of hospital capacity planning and optimisation. *Healthcare*, *10*(5), 826. https://doi.org/10.3390/healthcare10050826

Hüsch, P., Mott, G., MacColl, J., Nurse, J., Sullivan, J., Turner, S., & Pattnaik, N. (2024). *'Your data is stolen and encrypted': The ransomware victim experience.* University of Kent. https://kar.kent.ac.uk/106978/

Hwang, D., & Woodrum, R. R. (2020). Inadequacy of current solutions to address provider shortage. *CHEST*, *157*(5), 1055–1057. https://doi.org/10.1016/j.chest.2020.03.025

Jalali, M. S., Bruckes, M., Westmattelmann, D., & Schewe, G. (2020). Why employees (still) click on phishing links: Investigation in hospitals. *Journal of Medical Internet Research*, *22*(1), e16775. https://doi.org/10.2196/16775

Jennings, M. (2024, March 11). Medical records left on the lawn after Georgia hospital cleared out. *CBS*, *42*. https://www.cbs42.com/regional/georgia-news/medical-records-exposed-after-hospital-was-cleared-out/

Joseph, A. (2022, March 23). *"I fear the long-term effects": Before his death, a nurse warned of the pandemic's toll on health care workers.* STAT. https://www.statnews.com/2022/03/23/nurse-warned-of-pandemic-mental-toll-health-workers/

Kalmoe, M. C., Chapman, M. B., Gold, J. A., & Giedinghagen, A. M. (2019). Physician suicide: A call to action. *Missouri Medicine*, *116*(3), 211–216.

Kanokphanvanich, C., Rattanawong, W., & Vongmanee, V. (2022). A meta-analysis of health care supply chain performance for patient safety. *2022 14th International Conference on Software, Knowledge, Information Management and Applications (SKIMA)*, 13–20. https://doi.org/10.1109/SKIMA57145.2022.10029487

Khatri, N., Brown, G. D., & Hicks, L. L. (2009). From a blame culture to a just culture in health care. *Health Care Management Review*, *34*(4), 312–322. https://doi.org/10.1097/HMR.0b013e3181a3b709

Klindienst, J., Ayanian, S., Schlegelmilch, J., & Akselrod, H. (2022). Preparing for compounding crises: Staff shortages and cyber-attack vulnerability in the era of COVID-19. *Journal of Business Continuity & Emergency Planning*, *16*(2), 103–120.

Korte, M. (2020). The impact of the digital revolution on human brain and behavior: Where do we stand? *Dialogues in Clinical Neuroscience*, *22*(2), 101–111. https://doi.org/10.31887/DCNS.2020.22.2/mkorte

Ledoux, K. (2015). Understanding compassion fatigue: Understanding compassion. *Journal of Advanced Nursing*, *71*(9), 2041–2050. https://doi.org/10.1111/jan.12686

Leszczyna, R. (2021). Review of cybersecurity assessment methods: Applicability perspective. *Computers & Security*, *108*. https://doi.org/10.1016/j.cose.2021.102376

Li, L., Badgery-Parker, T., Merchant, A., Fitzpatrick, E., Raban, M. Z., Mumford, V., … & Westbrook, J. I. (2024). Paediatric medication incident reporting: A multicentre comparison study of medication errors identified at audit, detected by staff and reported to an incident system. *BMJ Quality & Safety*, *33*(10), 624–633.

Li, L., He, W., Xu, L., Ash, I., Anwar, M., & Yuan, X. (2019). Investigating the impact of cybersecurity policy awareness on employees' cybersecurity behavior. *International Journal of Information Management*, *45*, 13–24. https://doi.org/10.1016/j.ijinfomgt.2018.10.017

Linger, R., Lipson, H., McHugh, J., Mead, N., & Sledge, C. (2002). *Life-cycle models for survivable systems* (Technical Report CMU/SEI-2002-TR-026). Carnegie Mellon. https://apps.dtic.mil/sti/pdfs/ADA407793.pdf

Lucchini, A., Iozzo, P., & Bambi, S. (2020). Nursing workload in the COVID-19 era. *Intensive & Critical Care Nursing*, *61*, 102929. https://doi.org/10.1016/j.iccn.2020.102929

Lugto, K. (2024, May 29). *The rise of reverse social engineering scams*. MSN. https://www.msn.com/en-ph/news/money/the-rise-of-reverse-social-engineering-scams/ar-BB1nh1FX?apiversion=v2&noservercache=1&domshim=1&renderwebcomponents=1&wcseo=1&batchservertelemetry=1&noservertelemetry=1

Lusk, C., DeForest, E., Segarra, G., Neyens, D. M., Abernathy, J. H., & Catchpole, K. (2022). Reconsidering the application of systems thinking in healthcare: The RaDonda Vaught case. *British Journal of Anaesthesia*, *129*(3), e61–e62. https://doi.org/10.1016/j.bja.2022.05.023

Mayo Clinic. (n.d.). *Fatigue*. Mayo Clinic. Retrieved February 4, 2025, from https://www.mayoclinic.org/symptoms/fatigue/basics/definition/sym-20050894

Miller, M. (2022, December 28). *The mounting death toll of hospital cyberattacks—POLITICO*. POLITICO. https://www.politico.com/news/2022/12/28/cyberattacks-u-s-hospitals-00075638

Mishra, V. (2024, November 8). *Cyberattacks on healthcare: A global threat that can't be ignored | UN News*. United Nations. https://news.un.org/en/story/2024/11/1156751

Morey, D. A., Walli, P., Cassidy, K. S., Tewani, P. K., Reynolds, M. E., Malone, S., … & McGeorge, N. M. (2023, September). Towards joint activity design heuristics: Essentials for human-machine teaming. In *Proceedings of the Human Factors and Ergonomics Society Annual Meeting* (Vol. 67, No. 1, pp. 131–136). SAGE Publications.

Morrison, S. (2023, September 15). *The chaotic and cinematic MGM casino hack, explained*. Vox. https://www.vox.com/technology/2023/9/15/23875113/mgm-hack-casino-vishing-cybersecurity-ransomware

Moses, R. E., McNeese, L. G., Feld, L. D., & Feld, A. D. (2014). Social media in the healthcare setting: Benefits but also a minefield of compliance and other legal issues. *Official Journal of the American College of Gastroenterology | ACG*, *109*(8), 1128. https://doi.org/10.1038/ajg.2014.67

National Health Service. (2022, August). *SEIPS quick reference guide and work system explorer*. NHS England. https://www.england.nhs.uk/wp-content/uploads/2022/08/B1465-SEIPS-quick-reference-and-work-system-explorer-v1-FINAL.pdf

National Institute of Standards and Technology. (n.d.). *social engineering—Glossary | CSRC*. National Institute of Standards and Technology. Retrieved February 4, 2025, from https://csrc.nist.gov/glossary/term/social_engineering

Nifakos, S., Chandramouli, K., Nikolaou, C. K., Papachristou, P., Koch, S., Panaousis, E., & Bonacina, S. (2021). Influence of human factors on cyber security within healthcare organisations: A systematic review. *Sensors*, *21*(15). https://www.mdpi.com/1424-8220/21/15/5119

Office of Information Security. (2024, April 11). *Social Engineering attacks targeting the HPH sector* [PowerPoint]. https://www.hhs.gov/sites/default/files/social-engineering-targeting-the-hph-sector-tlpclear.pdf

Osuagwu, E. U., Chukwudebe, G. A., Salihu, T., & Chukwudebe, V. N. (2015). Mitigating social engineering for improved cybersecurity. *2015 International Conference on Cyberspace (CYBER-Abuja)*, 91–100. https://doi.org/10.1109/CYBER-Abuja.2015.7360515

Owens, B. (2020). How hospitals can protect themselves from cyber attack. *CMAJ : Canadian Medical Association Journal*, *192*(4), E101–E102. https://doi.org/10.1503/cmaj.1095841

Owens, J. (2007). Sleep loss and fatigue in healthcare professionals. *The Journal of Perinatal & Neonatal Nursing*, *21*(2), 92. https://doi.org/10.1097/01.JPN.0000270624.64584.9d

Petrosyan, A. (2024, December 10). *Weekly global cyberattacks by industry 2023*. Statista. https://www.statista.com/statistics/1377217/average-weekly-number-attacks-global-by-industry/

Porter, J., Boyd, C., Skandari, M. R., & Laiteerapong, N. (2023). Revisiting the time needed to provide adult primary care. *Journal of General Internal Medicine*, *38*(1), 147–155. https://doi.org/10.1007/s11606-022-07707-x

Priestman, W., Anstis, T., Sebire, I. G., Sridharan, S., & Sebire, N. J. (2019). Phishing in healthcare organisations: Threats, mitigation and approaches. *BMJ Health & Care Informatics*, *26*(1), e100031. https://doi.org/10.1136/bmjhci-2019-100031

Queensland Government. (2022, May 11). *Social media for business* [Text]. Queensland Government; CorporateName=The State of Queensland. https://www.business.qld.gov.au/running-business/marketing-sales/marketing/websites-social-media/social-media

Rathore, S., Sharma, P. K., Loia, V., Jeong, Y.-S., & Park, J. H. (2017). Social network security: Issues, challenges, threats, and solutions. *Information Sciences*, *421*, 43–69. https://doi.org/10.1016/j.ins.2017.08.063

Ribeiro, A. (2024, September 17). CPR data reports 32% rise this year, as global healthcare sector faces surge in cyberattacks. *Industrial Cyber*. https://industrialcyber.co/medical/cpr-data-reports-32-rise-this-year-as-global-healthcare-sector-faces-surge-in-cyberattacks/

Rigas, N., Kyritsis, Z., Dagla, M., Soldatou, A., Orovou, E., Tzitiridou-Chatzopoulou, M., Eskitzis, P., & Antoniou, E. (2024). Post-traumatic stress disorder symptoms among pediatric healthcare workers. *Nursing Reports*, *14*(1), 115–127. https://doi.org/10.3390/nursrep14010010

Rink, L. C., Oyesanya, T. O., Adair, K. C., Humphreys, J. C., Silva, S. G., & Sexton, J. B. (2023). Stressors among healthcare workers: A summative content analysis. *Global Qualitative Nursing Research*, *10*, 23333936231161127. https://doi.org/10.1177/23333936231161127

Ružić, D., & Jovanović, N. (2023). Compromising personal safety on social media: The fine line between technical limitations and personal choices. In M. Boskovic, G. Misev, & N. Putnik (Eds.), *Analyzing new forms of social disorders in modern virtual environments* (pp. 42–65). IGI Global Scientific Publishing. https://doi.org/10.4018/978-1-6684-5760-3.ch003

Samad, A. (2024). *Social engineering in cybersecurity: Prevention and mitigation strategies (13857)*. EasyChair Preprint.

Schein, E. H. (1984, January 15). *Coming to a new awareness of organizational culture.* MIT Sloan Management Review. https://sloanreview.mit.edu/article/coming-to-a-new-awareness-of-organizational-culture/

Selwyn, N., & Stirling, E. (2016). Social media and education … now the dust has settled. *Learning, Media and Technology, 41*(1), 1–5. https://doi.org/10.1080/17439884.2015.1115769

Skyvell Nilsson, M., Törner, M., & Pousette, A. (2018). Professional culture, information security and healthcare quality—an interview study of physicians' and nurses' perspectives on value conflicts in the use of electronic medical records. *Safety in Health, 4*(1), 11. https://doi.org/10.1186/s40886-018-0078-9

Staal, M. A. (2004). *Stress, cognition, and human performance: A literature review and conceptual framework* (Technical Memorandum NASA/TM-2004-212824). Ames Research Center. https://ntrs.nasa.gov/api/citations/20060017835/downloads/20060017835.pdf

Stanton, B., Theofanos, M. F., Prettyman, S. S., & Furman, S. (2016). Security fatigue. *IT Professional, 18*(5), 26–32. https://doi.org/10.1109/MITP.2016.84

St. Thomas University. (2023, November 26). *The power of social media marketing.* St. Thomas University. https://online.stu.edu/degrees/business/mba/marketing/the-power-of-social-media/

Syafitri, W., Shukur, Z., Mokhtar, U. A., Sulaiman, R., & Ibrahim, M. A. (2022). Social engineering attacks prevention: A systematic literature review. *IEEE Access, 10*, 39325–39343. IEEE Access. https://doi.org/10.1109/ACCESS.2022.3162594

The Joint Commission. (2023). *Preserving patient safety after a cyberattack* [Sentinel Event Alert]. The Joint Commission.

The Physicians Foundation. (2024). *2024 survey of America's current and future physicians: Examining physician, resident, and student wellbeing and impact of the current healthcare landscape.* The Physicians Foundation. https://physiciansfoundation.org/wp-content/uploads/2024-Survey-of-Americas-Current-and-Future-Physicians.pdf

Tischer, M., Durumeric, Z., Foster, S., Duan, S., Mori, A., Bursztein, E., & Bailey, M. (2016). Users really do plug in USB drives they find. *2016 IEEE symposium on security and privacy (SP)*, 306–319. https://doi.org/10.1109/SP.2016.26

University of Pennsylvania. (n.d.). *Where did all the nurses go?* PennNursing. Retrieved February 4, 2025, from https://www.nursing.upenn.edu/nhhc/workforce-issues/where-did-all-the-nurses-go/

Venkatanathan, J., Kostakos, V., Karapanos, E., & Gonçalves, J. (2014). Online disclosure of personally identifiable information with strangers: Effects of public and private sharing. *Interacting with Computers, 26*(6), 614–626. https://doi.org/10.1093/iwc/iwt058

Verbeke, W., Volgering, M., & Hessels, M. (1998). Exploring the conceptual expansion within the field of organizational behaviour: Organizational climate and organizational culture. *Journal of Management Studies, 35*(3), 303–329. https://doi.org/10.1111/1467-6486.00095

Wagner, M. (2014, June 19). Justina Pelletier's parents win custody; 16-year-old finally goes home after 16 months in Massachusetts state care. *New York Daily News.* https://www.nydailynews.com/2014/06/19/justina-pelletiers-parents-win-custody-16-year-old-finally-goes-home-after-16-months-in-massachusetts-state-care/

Wang, D., Wang, X., & Xia, N. (2018). How safety-related stress affects workers' safety behavior: The moderating role of psychological capital. *Safety Science, 103*, 247–259. https://doi.org/10.1016/j.ssci.2017.11.020

White, G. R. T., Allen, D. R., Samuel, D. A., Taylor, D., Thomas, D. R., & Jones, P. (2022). The ecosystem of UK social entrepreneurship: A meta-analysis of contemporary studies. In D. G. Pickernell, M. Battisti, Z. Dann, & C. Ekinsmyth (Eds.), *Disadvantaged entrepreneurship and the entrepreneurial ecosystem* (Vol. 14, pp. 1–8). Emerald Publishing Limited.

Williamson, A., Lombardi, D. A., Folkard, S., Stutts, J., Courtney, T. K., & Connor, J. L. (2011). The link between fatigue and safety. *Accident Analysis & Prevention*, *43*(2), 498–515. https://doi.org/10.1016/j.aap.2009.11.011

Wolf, J. M. (2017). The multipurpose tool of social media: Applications for scientists, science communicators, and educators. *Clinical Microbiology Newsletter*, *39*(10), 75–79. https://doi.org/10.1016/j.clinmicnews.2017.04.003

Zulkurnain, A. U., Hamidy, A. K. B. K., Husain, A. B., & Chizari, H. (2015). Social engineering attack mitigation. Int. J. Math. Comput. Sci, *1*(4), 188-198.

8 Machine Learning for Device Identification with Application in Aviation Cybersecurity

Yongxin Liu, Jian Wang, and Houbing Song

As a rapidly evolving field, the Internet of Things (IoT) connects smart devices equipped with sensors, onboard data processing, and communication capabilities to enable automation across industries (Jeschke et al., 2017). In aviation, IoT plays a crucial role in flight operations, air traffic control (ATC), and maintenance, enabling real-time monitoring and decision-making (Dartmann et al., 2019). Automated aircraft tracking via Automatic Dependent Surveillance–Broadcast (ADS-B), predictive maintenance using connected sensors, and unmanned aerial systems (UAS) integration rely heavily on IoT technologies. However, ensuring the security, reliability, and human interpretability of these systems remains a critical challenge.

Despite its benefits, IoT introduces security risks due to increased connectivity (Butun et al., 2019; Song et al., 2017). Rogue IoT devices pose a significant threat, especially in safety-critical domains like aviation, where compromised devices could manipulate flight data, disrupt ATC operations, or execute cyber-physical attacks. These attacks range from identity spoofing and unauthorized control access to distributed denial of service (DDoS) attacks and data breaches (Butun et al., 2019). Addressing these risks requires robust detection and identification mechanisms. Machine learning (ML), as a data-driven approach, offers significant potential to enhance IoT security by detecting anomalies, classifying rogue devices, and mitigating cyber threats.

We present a comprehensive survey of ML-based techniques for detecting and identifying rogue IoT devices. Additionally, we examine enabling technologies such as statistical learning, feature engineering, digital signal processing, and deep learning. Core ML tools—including continual learning, unsupervised learning, explainable artificial intelligence (XAI), and anomaly detection—are explored to assess their applicability in aviation cybersecurity.

Figure 8.1 illustrates the attack chain in IoT systems. By integrating ML-driven solutions with human-centric design principles, aviation IoT security can move toward a more resilient and interpretable framework.

THREAT MODE AND DETECTION FRAMEWORK OF ROGUE DEVICES IN IOT

The nature of IoT allows sharing information and flexible collaboration among a large number of heterogeneous devices. This standard, on the one hand, cultivates a

 DOI: 10.1201/9781003413707-9

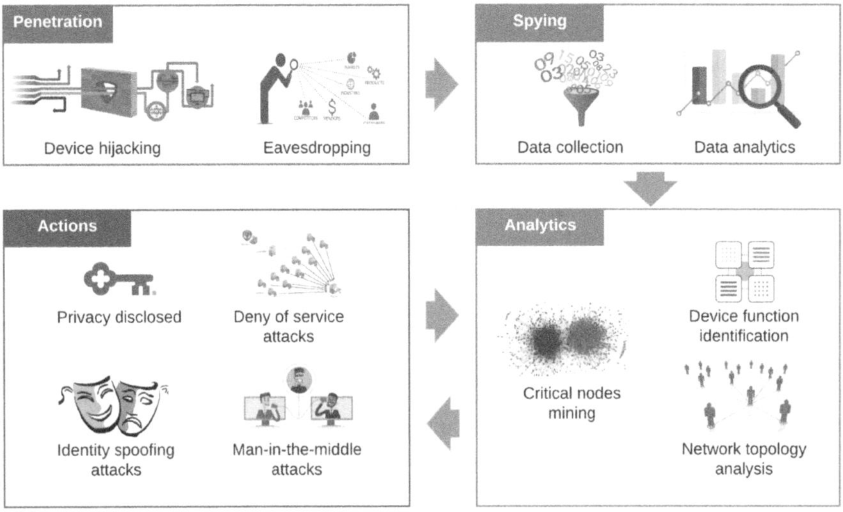

FIGURE 8.1 Attack chain in the IoT.

large open system with low entry restrictions. On the other hand, this open system and the limited capacity of IoT devices together allow adversaries to conduct rogue activities with great convenience (Wang et al., 2018).

IoT adversaries use two attack modes: passive and proactive. Passive attacks involve stealthy monitoring of device communications to gather intelligence for future exploits. In contrast, proactive attacks actively degrade system performance or steal sensitive information.

In practical attacks, proactive and passive attacks are combined. A typical attack chain to IoT systems is shown in Figure 8.1. We divided the whole attack chain into five stages, as follows:

1. *Penetration:* In this stage, the rogue IoT device tries to eavesdrop communication channels or attain the control privilege of vulnerable peers in the IoT for further actions.
2. *Spying:* In this stage, the rogue IoT device will capture and store the ongoing activities of the IoT by exploiting penetrated devices as Trojans.
3. *Data analytics:* The rogue IoT devices conduct a series of off-line data mining to model the behaviors and evaluate the vulnerabilities of the IoT from multiple perspectives.
4. *Attacking:* In this stage, the adversaries perform strategic planning and wait for the best time to minimize their risk while maximizing the rewards.
5. *Attack:* In this stage, prevalent attacks are conducted to the target.

Among these stages, passive and proactive attacks are combined in the penetration stage.

If we can prevent the adversaries from successfully impersonating legitimate devices in the first stage (penetration) or can identify hijacked devices in the second stage (spying), the whole attack chain can be destroyed.

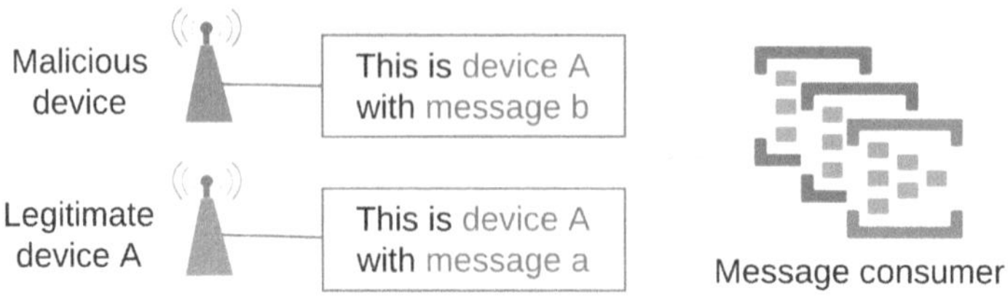

FIGURE 8.2 Threat mode of rogue IoT devices.

We denote rogue IoT devices as unknown IoT devices impersonating some known devices or hijacked legitimate devices to access sensitive resources in target systems. The attack schemes of rogue IoT devices are depicted in Figure 8.2. As depicted, both legitimate devices and unknown devices can be exploited by adversaries. Non-cryptographic mechanisms need to be capable of (a) verifying the identity of devices and (b) identifying hijacked devices or devices with abnormal behaviors. Both of the two goals require data modeling and ML techniques; in the following section, we cover the basic data processing pipeline of physical-layer device identification.

FEATURE-BASED STATISTICAL LEARNING APPROACHES

The feature-based approach requires human efforts to decide the relevant features to uniquely distinguish a transmitter from another.

Unclonable Features

Manufacturing imperfections in wireless devices introduce subtle, unintentional variations in their radio frequency (RF) frontends. While these variations do not degrade communication performance, they create unique, device-specific signatures that can be leveraged for identification. Since users cannot modify or replicate these hardware-induced characteristics, they are termed Physical Unclonable Features (PUFs) (Chatterjee et al., 2018; Herder et al., 2014).

Error/Noise Patterns

Received signals often deviate from theoretical expectations due to hardware imperfections. These deviations, typically treated as noise, can serve as distinguishing features for device identification. Azarmehr et al. (2017) and Polak and Goeckel (2015b) demonstrated that phase errors in phase lock loop (PLL) circuits can uniquely characterize transmitters, even in low- Signal-to-Noise Ratio (SNR) (Signal) environments. Zhuang et al. (2018) extracted error vectors by comparing received I/Q signals with expected templates, integrating statistical and time-frequency features for fingerprinting. Their work also exemplifies data fusion techniques (see Section Hybrid Feature Fusion). Notably, similar methods are widely used in digital forensics for camera model identification (Cozzolino & Verdoliva, 2020).

Peng et al. (2016, 2018, 2019) employed differential constellation trace figure (DCTF), analyzing carrier frequency offset, modulation offset, and I/Q imbalances in Zigbee devices. Their lightweight classifier dynamically adjusts feature weights based on SNR conditions, yielding robust performance across 54 tested Zigbee devices. Their approach mirrors k-nearest neighbor (k-NN) behavior in classification.

Persistent Patterns

Persistent pattern-based recognition leverages statistical features extracted from signal slices to identify transmitters, also referred to as modulation domain recognition. A well-known method, RF-distinctive native attributive (RF-DNA), uses statistical metrics from signal sub-regions as fingerprints (Cobb et al., 2010).

Bihl et al. (2016), Dubendorfer et al. (2012), and Ramsey et al. (2015) analyzed Wireless Personal Area Network (WPAN) signal preambles, extracting variance, skewness, and kurtosis from signal sub-regions (bins) to generate unique signatures. Their experiments suggest that these simple statistical features can effectively distinguish transmitters. Similarly, Harmer et al. (2011) demonstrated that RF-DNA can also derive frequency domain fingerprints.

Transient Patterns

Unlike persistent statistical features, transient PUFs capture dynamic characteristics of RF transmitters, making them more resistant to forgery in wireless channels (Danev et al., 2010). These transient signals typically occur at the beginning and end of wireless packet transmissions, as illustrated in Figure 8.3.

Polak and Goeckel (2015a) leveraged non-linear in-band distortion and spectral regrowth, likely introduced by power amplifiers, to differentiate masquerade emitters. Their approach is resilient against adversaries attempting to tamper with baseband waveform symbols. Similarly, Köse et al. (2019) analyzed the energy spectrum of the transient amplitude envelopes on the transmitter to classify eight different devices, demonstrating that the frequency domain features outperform the time domain. Tu et al. (2019) and Zhang et al. (2018) further utilized time domain statistical metrics and wavelet transforms of transient signals as RF fingerprints.

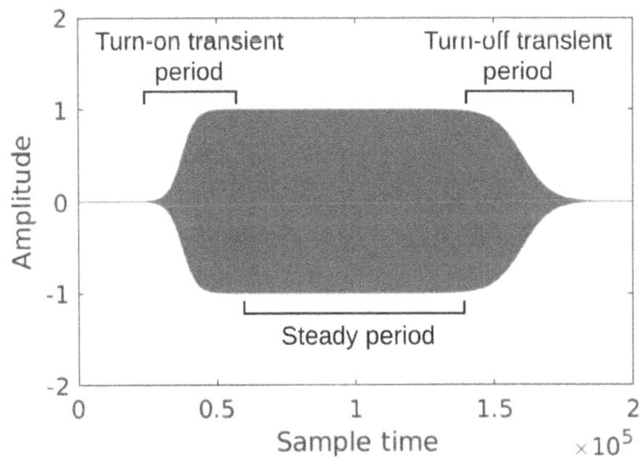

FIGURE 8.3 Transient periods during wireless communication.

Expanding on this, Shi et al. (2015) applied multifractal theory to Fourier energy spectra of WiFi transceivers' transient signals, achieving superior identification accuracy compared to conventional time domain and wavelet-based methods. However, they noted that transmitters from the same manufacturer are more challenging to distinguish. Ali et al. (2019) focused on Bluetooth devices, extracting 13 time-frequency domain features using the Hilbert-Huang spectrum. Their decision tree-based classification demonstrated the efficacy of well-designed transient fingerprints in device identification.

Hybrid Feature Fusion

To enhance distinguishability, many researchers transform signals into alternative domains, particularly the time-frequency domain. Chen et al. (2017) combined short-time Fourier transform (STFT) with support vector machine (SVM) to classify four different transceivers, while Reising et al. (2015) applied the discrete Gabor transform (a Gaussian-windowed STFT) for a similar purpose. Li et al. (2018) and Wu et al. (2019) utilized wavelet transforms alongside SVM and probabilistic neural networks, incorporating principle component analysis (PCA) for dimensionality reduction to improve efficiency. Sun et al. (2016) introduced a frequency-based method combining PCA and SVM for Global System for Mobile Communication (GSM) emitter identification, comparing its effectiveness against the Hilbert-Huang transform (Yuan et al., 2014). Satija et al. (2018) further demonstrated that variational mode decomposition (VMD) could outperform traditional empirical mode decomposition (EMD) in relay scenarios.

Beyond time-frequency analysis, other domains also prove useful for emitter identification. Kim et al. (2008) leveraged cyclic frequency metrics to differentiate WiFi transmitters. Wang et al. (2017) combined bispectrum-based energy entropy and color moments with SVM, showing that higher-order statistics improve identification in low-SNR environments. However, Lei (2016) found that square integral bispectra (SIB) is more noise-resistant while preserving the same discriminative information as the bispectrum.

Sun et al. (2017) used singular values of axial integrated Wigner bispectrum (AIWB) to distinguish spoofed signals from genuine ones in Global Navigation Satellite System (GNSS). Meanwhile, Deng et al. (2017) employed multidimensional permutation entropy to capture the complex variations in wireless signal patterns for transmitter profiling.

Implications to Human-Machine Interface Design

Feature-based statistical learning enhances the security of human-machine interface (HMI) systems by enabling real-time identification of rogue IoT devices through physical-layer characteristics. Since traditional security methods often rely on cryptographic authentication, which can be vulnerable to key disclosure or computational inefficiencies, statistical learning provides an alternative by leveraging inherent device fingerprints such as error patterns, transient signals, and modulation variations. This ensures that HMIs interacting with IoT systems can verify the

TABLE 8.1
A Brief Comparison of Classifiers in Deployable Wireless Transmitter Identification Systems

Approach	Application Overhead	Continual Learning	Anomaly Detection
k-NN	Depends on the size of fingerprint library	Natively supported	Clustering or statistical models
SVM	Depends on the number of feature dimensions	Regularization	One-class SVM
Random forest	Depends on the number of decision trees	Knowledge replay	Isolation forest
Neural network	Depends on structural complexity	Section	Section

authenticity of connected devices without requiring additional user input, reducing cognitive load while maintaining security.

For effective integration into HMI design, these statistical learning models must be interpretable and user-friendly. Operators managing secured systems need clear, actionable insights rather than raw signal processing outputs. This requires designing HMIs that (Table 8.1) present risk assessments and device classifications in an intuitive manner, possibly through visual alerts, confidence scores, and anomaly explanations. Additionally, real-time feedback mechanisms can help users understand why a device is flagged as suspicious, supporting decision-making in high-stakes environments like aviation or industrial control systems.

Lastly, the adaptability of statistical learning approaches plays a crucial role in dynamic HMI environments. As IoT ecosystems evolve, new device types and attack patterns emerge, requiring continual model updates without overwhelming end users. Implementing adaptive interfaces that adjust security recommendations based on evolving threat landscapes ensures that HMIs remain both effective and user-centric. By integrating ML-based anomaly detection with interactive HMI components, secured systems can strike a balance between automation and human oversight, enhancing both usability and resilience against cyber threats.

DEEP LEARNING FOR DEVICE IDENTIFICATION

A key limitation of device-specific feature-based statistical learning is the need for manual selection of relevant transforms or features. To address this, researchers have increasingly adopted deep learning techniques to automatically extract latent features from received signals, enabling more efficient fingerprinting of RF transmitters. This section reviews deep learning-based wireless device identification methods and examines the persistent challenges associated with deep learning and neural networks in this domain.

A typical convolutional neural network (CNN)-based wireless device identification system is shown in Figure 8.4. CNNs extract latent features using convolutional layers and classify them through fully connected dense layers. Unlike conventional

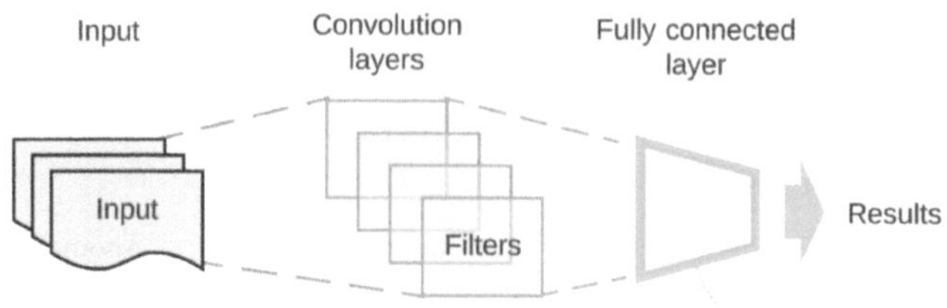

FIGURE 8.4 Typical architecture of deep neural network.

fully connected neural networks, CNNs employ filters (kernels) that reduce parameter count while enhancing feature extraction, significantly minimizing the need for manual feature engineering.

Case Studies and Comparisons

Deep learning has shown strong performance in this domain. Yu, Hu, Zhou, et al. (2019) integrated autoencoders and CNNs for signal denoising and emitter identification, achieving robust results even in low-SNR conditions. Huang et al. (2017) used a stacked denoising autoencoder to learn low-dimensional representations of radio signals. Merchant et al. (2018) trained a deep network using differential errors between reconstructed and received signals to classify Zigbee transceivers, eliminating manual feature selection.

Additionally, Baldini et al. (2018) found that wavelet features outperform STFT features for radiometric fingerprinting.

CNNs have demonstrated high accuracy with raw I/Q baseband data. Riyaz et al. (2018) optimized a CNN for emitter classification in 802.11AC channels, outperforming SVM and logistic regression. Morin et al. (2019) trained neural networks on raw I/Q samples, showing CNN's potential despite performance degradation due to emitter movement. Roy, Mukherjee, Chatterjee, and Pasiliao (2020) found that deep neural networks (DNNs) outperform CNNs in classifying in-phase and quadrature samples. Yu, Hu, Li, and Peng (2019) refined CNN-based classification by extracting 1-D regions of interest (ROI) from Zigbee preambles, resampling signals at varying rates before classification. Preventing models from learning trivial features like protocol identifiers remains a key challenge.

Deep learning requires large datasets. Oyedare and Park (n.d.) estimated dataset size needs for CNN-based classification, concluding it should be 10,000 to 30,000 times the number of transmitters. Network architecture also plays a critical role—Agadakos et al. (2019) demonstrated that complex-valued deep networks outperform real-valued ones for transmitter identification in WiFi and ADS-B datasets.

Open Issues in Deep Learning for Wireless Transmitter Identification

The DNN is becoming a promising technology in this domain. However, as in other domains, DNN encounters several problems. Although researches in wireless transmitter identification rarely cover the issues, we briefly review its current state and solutions.

Hyperparameter Searching

Hyperparameter tuning significantly impacts DNN performance, influencing factors such as learning rate, mini-batch size, and dropout rate. Jafari et al. (2018) compared CNN, DNN, and long short-term memory (LSTM) in wireless transmitter identification using I/Q baseband signals, finding that CNN performed best, followed by DNN and LSTM, with hyperparameter choices playing a crucial role. However, optimizing hyperparameters is computationally intensive as their effects are unpredictable before training.

Several search strategies have been developed to address this challenge. The grid search systematically evaluates all combinations of parameters, but is inefficient for large search spaces. Random search improves efficiency by randomly sampling hyperparameters and performs well when only a few parameters significantly impact performance. Prediction-based methods, such as Bayesian optimization (Bergstra et al., 2011; Pelikan et al., 1999), model network performance based on initial trials to identify promising configurations. Evolutionary algorithms, including genetic algorithms (Young et al., 2015), explore the search space heuristically but lack guarantees of outperforming prediction-based methods. Many of these strategies are now integrated into AutoML platforms, with Truong et al. (2019) providing a comparative analysis.

Open Set Recognition

Deep learning models typically recognize only pre-trained classes, making them ineffective for identifying previously unseen transmitters. Wong et al. (2018) addressed this limitation using a semi-supervised approach, training a CNN on known transmitters before removing the softmax layer to transform raw signals into non-linear feature representations. These features were then clustered using Density-Based Spatial Clustering of Applications with Noise (DBSCAN), demonstrating the potential to detect untrained emitters, although the approach remains in its early stages. Similarly, Youssef et al. (2018) proposed incremental learning to allow neural networks to classify newly registered transmitters over time.

This challenge, categorized as open set recognition (Bendale & Boult, 2016; Scheirer et al., 2012), aligns with anomaly detection. Roy, Mukherjee, Chatterjee, Blasch, and Pasiliao (2020) employed generative adversarial networks (GANs) to generate synthetic signals, using the discriminator to distinguish abnormal inputs. Shi et al. (2019) introduced two alternative methods: transforming signals into feature vectors and detecting outliers using Mahalanobis distance, or applying k-means clustering to separate unknown signals from known classes. These approaches aim to improve the adaptability of deep learning models in dynamic environments.

Continual Learning

DNNs in practical applications must adapt to new tasks over time. Wireless device identification systems, for example, need to recognize newly encountered devices throughout their lifecycle. This capability falls under lifelong learning, which includes *transfer learning (TL)* and *continuous learning (CL)*. TL involves pre-training models in a controlled environment and fine-tuning them for deployment (Tan et al., 2018). In contrast, CL enables models to incrementally learn from new data while retaining previously acquired knowledge (Parisi et al., 2019). Unlike TL, CL prevents *catastrophic forgetting*, where a model loses prior knowledge when trained on new data. TL is suitable for initial deployments, while CL is essential for continuous updates.

Continual learning in DNNs is achieved through three main strategies: knowledge replay, regularization, and dynamic network expansion. Knowledge replay mitigates forgetting by retraining on past data, though storing large datasets is resource-intensive; to address this, Shin et al. (2017) used GANs to generate synthetic samples, reducing memory overhead. Regularization preserves prior knowledge by constraining weight updates; global regularization limits changes across the network but restricts adaptability, while local methods like elastic weight consolidation (EWC) (Kirkpatrick et al., 2017) selectively protect critical connections while allowing flexibility in non-essential ones. Dynamic network expansion prevents forgetting by freezing existing parameters and growing the network for new tasks, as seen in dynamic expanding network (DEN) (Yoon et al., 2017), which identifies and duplicates relevant neurons to accommodate additional learning. Each approach balances memory efficiency, adaptability, and computational cost, making the choice dependent on application requirements.

Implications on Human-Machine Interface Design

Deep learning offers powerful capabilities for device identification in secured systems by automatically extracting complex patterns from raw signals, reducing reliance on manually engineered features. This significantly enhances the accuracy and adaptability of HMI systems by enabling automated threat detection with minimal human intervention. Unlike feature-based statistical learning, which depends on predefined characteristics like error patterns and transient signals, deep learning models continuously evolve and improve as they are exposed to more data. This adaptability is crucial in high-security environments where IoT threats evolve rapidly, ensuring that HMIs provide up-to-date security insights without requiring constant manual tuning.

However, deep learning introduces challenges in HMI design, particularly regarding interpretability. Statistical learning models offer more explainable decision-making by relying on human-recognizable features, whereas deep learning operates as a "black box," making it difficult for users to understand why a device is flagged as rogue. This lack of transparency can reduce user trust and hinder response efficiency, especially in high-stakes settings like aviation security or critical infrastructure protection. To mitigate this, HMIs must incorporate XAI techniques that translate deep learning outputs into human-interpretable formats, such as visual heatmaps, decision rationales, or confidence scores. Enhancing model transparency ensures that security operators can make informed decisions rather than blindly relying on system recommendations.

Additionally, deep learning-driven HMIs require a more robust computational infrastructure, which can affect system responsiveness and real-time usability. While statistical learning approaches are often lightweight and suitable for low-power IoT environments, deep learning demands significant processing power and data storage, potentially leading to latency issues. To balance security with efficiency, HMI design must incorporate hybrid approaches—leveraging statistical learning for quick anomaly detection while using deep learning for high-confidence verifications when necessary. By integrating these two methods within an adaptive HMI framework, secured systems can achieve both high accuracy and user-friendly operation, ensuring that security personnel remain engaged and informed without being overwhelmed by opaque or slow decision-making processes.

ABNORMAL DEVICE IDENTIFICATION

Intrusion detection systems (IDSs) play a crucial role in breaking the IoT attack chain (Figure 8.1) by identifying ongoing threats and compromised devices. Unlike device-level identification schemes, IDS typically rely on unsupervised learning since supervised methods require labeled data, which is difficult to obtain due to the diverse deployment environments of IoT systems. IDS generally build reference models from normal device behavior and detect deviations indicative of anomalies.

Statistical Modeling

Statistical modeling methods assess whether a device operates abnormally by analyzing behavioral patterns. Alipour et al. (2015) used a Markov chain model to evaluate IEEE 802.11 device state transitions, detecting compromised devices based on transition probabilities. Similarly, Sehatbakhsh et al. (2018) modeled electromagnetic (EM) harmonics of medical IoT devices, assuming that rogue shellcode execution would alter their expected signal distributions. However, these approaches require manual selection of key features, which can limit scalability.

Reconstruction Approaches

Reconstruction approaches rely on deep autoencoders to learn normal device behavior and flag deviations. A deep autoencoder compresses input data into a low-dimensional representation before reconstructing it. If an input deviates significantly from normal patterns, the reconstruction error will be high, signaling an anomaly. Thing (2017) used 155 packet features in stacked autoencoders trained on normal operational data for anomaly detection. Similar methods were explored in Luo and Nagarajan (2018) and Sridharan et al. (2018), where autoencoders detected anomalies based on packet analysis, followed by secondary classification of intrusion types. Meidan et al. (2018) demonstrated that deep autoencoders outperform one-class SVM (Chen et al., 2001), Isolation Forest (Ding & Fei, 2013), and local outlier factor (Kriegel et al., 2009) in anomaly detection.

Prediction Approaches

Prediction approaches leverage time-series models to anticipate device behavior and detect deviations. Parwez et al. (2017) applied CNN-based predictors to cellular network usage patterns, showing that prediction errors increase when anomalies occur. Similarly, Wei and Kim (2012) employed autoregression to capture normal traffic variations, while recent works (Rajendran et al., 2018; Tandiya et al., 2018) integrated deep autoencoders with LSTM networks to generate predictive models. These methods analyze deviations in time-frequency video streams, even identifying attack types based on spatial error distributions.

Open Issues

Despite these advancements, IDS solutions require effective post-processing of error vectors to balance detection accuracy, false positives, and response latency. The lack of standardized benchmarks and generalizable detection frameworks remains an open issue, highlighting the need for further research in optimizing anomaly detection and response mechanisms.

Implications on Human-Machine Interface Design

Abnormal device identification plays a critical role in secured systems by detecting deviations from expected device behavior, offering a broader defense against unknown threats compared to both statistical and deep learning-based identification. This anomaly-driven approach is particularly useful in HMI design, as it enables real-time monitoring of device activity without requiring prior knowledge of specific attack signatures. Unlike statistical learning, which relies on predefined features, and deep learning, which classifies devices based on learned patterns, abnormal device identification continuously adapts to emerging threats by analyzing behavioral inconsistencies. This provides HMIs with a proactive defense mechanism, allowing operators to focus on assessing anomalies rather than manually updating detection models.

However, anomaly detection also presents unique challenges for human factors. Because it flags deviations rather than explicitly identifying rogue devices, false positives can overwhelm users, leading to alert fatigue. In contrast, statistical learning offers more structured, rule-based identification, and deep learning provides confidence-based classifications, making their outputs more deterministic. To enhance usability, HMIs must incorporate adaptive alerting mechanisms that prioritize critical anomalies while filtering out benign deviations. Implementing tiered alerts—such as low-risk deviations requiring monitoring and high-risk anomalies demanding immediate intervention—can prevent cognitive overload and ensure efficient decision-making in secured environments like aviation security or industrial IoT networks.

Additionally, the interpretability of abnormal device identification is often less straightforward than statistical learning but more transparent than deep learning. Statistical learning provides clear, explainable rules for device classification, while deep learning often functions as a "black box." Anomaly detection falls in between, requiring HMIs to visualize behavioral deviations in an understandable format. Techniques such as time-series anomaly heatmaps, deviation trend analysis, and contextual risk assessments can improve situational awareness for security personnel. By integrating abnormal device detection with both statistical and deep learning models, HMIs can offer a comprehensive security solution that balances proactive anomaly detection, explainability, and operational efficiency, ensuring that human operators can effectively respond to threats without excessive reliance on automated systems.

APPLICATION OF DEVICE IDENTIFICATION IN AVIATION

Aviation traffic surveillance has gained increasing attention due to its critical role in safety and security. A breach in these systems can lead to catastrophic failures, while advancements in aviation IoT are also driving innovations in vehicular networks and UAS. Aviation IoT primarily enables real-time broadcasting of position, velocity, and heading to ATC and nearby aircraft, with ADS-B over the 1090ES data link being the most widely used standard (Sun, 2017).

ADS-B is essential for modern aviation safety, as ATC and pilots rely on its data for collision avoidance and control decisions (Strohmeier et al., 2014). Crowdsourced reports indicate that billions of ADS-B messages are transmitted daily, with usage continuing to rise (Schäfer et al., 2018). However, ADS-B lacks cryptographic authentication, making it vulnerable to adversarial exploitation (Costin & Francillon, 2012).

AVIATION-SPECIFIC THREAT MODES AND COUNTERMEASURES

The goal of the malicious is to provide misleading information to the ground-based ATC systems as depicted in Figure 8.5. In general, we can categorize the major threats into three modes:

Ghost aircraft: The malicious broadcasts fake ADS-B messages to convince the ATC of a non-exist aircraft.

Legitimate transmitter hijacking: The malicious penetrates into the control system of an aircraft and then let the transmitters to broadcast misleading messages. In this scheme, the malicious does not modify *ICAO ID* or *Call Sign* (Sun, 2017) but broadcasts erroneous information to other receivers.

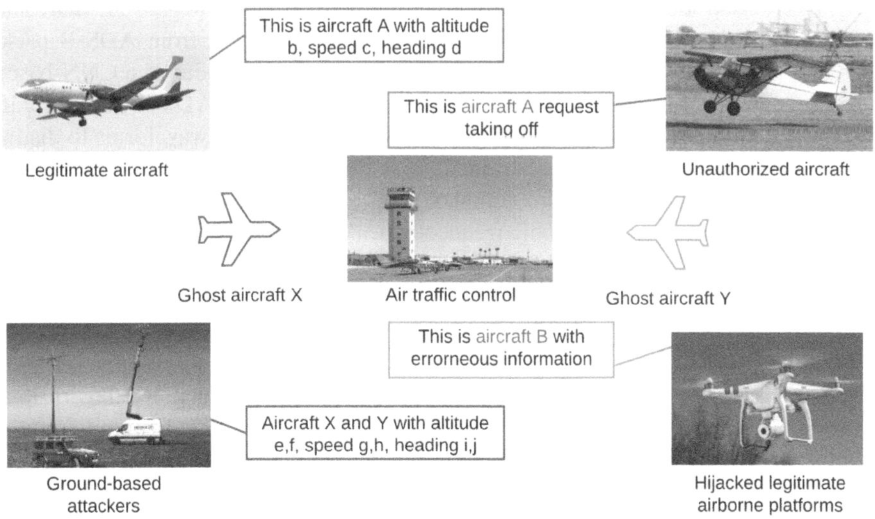

FIGURE 8.5 Attacks in air traffic surveillance system.

Identify spoofing: The malicious records the raw I/Q data of legitimate messages and uses preambles and identifiers of recorded messages to construct their own messages. In this scheme, the malicious provides accurate flight status information to other receivers, but they hide their real identity.

Since millions of devices are installed, it's not feasible to upgrade existing devices for authentication functionalities in the near future. Therefore, it's widely believed that non-cryptographic solutions are preferable to guarantee trustworthiness in the ADS-B systems. The non-cryptographic identification and verification in the ADS-B systems are in two folds:

- *Device-level verification*: To verify that a message is transmitted by a known RF transmitter (a.k.a. ADS-B transponder). A great majority of technologies in wireless transmitter identification can be applied to this topic.
- *Aircraft-level verification*: To verify that a message is transmitted from a known aircraft. Corresponding technologies in this topic are less commonly used in wireless transmitter identification.

Radiometric Fingerprinting Approaches

Leonardi and Di Fausto (2018a) demonstrated that signal phase features are more effective than frequency deviation and message patterns for distinguishing ADS-B transponders. Leonardi et al. (2017) applied instantaneous phase variations of intermediate frequency signals to a simple two-layer neural network, successfully identifying seven aircraft. Further tests (Leonardi & Di Fausto, 2018b) revealed that conventional statistical methods (K-NN and PCA) outperform shallow neural networks when analyzing sequences of received messages, while single-message classification remains unreliable.

Phase-based identification is feasible on low-cost receivers (e.g., RTL-SDR and HackRF). Ying et al. (2019) extracted per-sample phase data from ADS-B packets and used DNNs to detect spoofing. Chen et al. (2019) introduced a CNN-based classifier with a residual-reception structure, leveraging raw ADS-B and Aircraft Communications Addressing and Reporting System (ACARS) waveforms to distinguish RF transmitters. Their pre-trained model, trained on 50 labeled planes, demonstrated effective TL for more complex tasks. Moser et al. (2016) refined phase error measurements and harmonic frequency peaks to create robust RF fingerprints, showing resilience even in low-SNR conditions.

Context-Based Verification

Context-based approaches analyze transponder behavior rather than RF fingerprints. Strohmeier (2017) and Strohmeier and Martinovic (2015) utilized timing features between ADS-B messages to fingerprint transponders, enabling clustering without complex signal processing. Their method applies directly to open-source ADS-B datasets (e.g., OpenSky, FlightRadar24, and ADS-B Exchange) but does not identify individual transponders. In contrast, Habler and Shabtai (2018) proposed an IDS using an LSTM-enabled autoencoder, detecting spoofed or hijacked broadcasts by measuring reconstruction errors.

Flight Dynamics-Based Approaches

Flight dynamics provide additional verification by ensuring messages originate from aircraft rather than ground-based spoofers. These methods use location and velocity consistency checks.

Location-Based Methods

These methods assume that fake transmitters are at different locations than real aircraft, verifying messages by comparing reported and estimated positions. Monteiro, Barreto, Kacem, Carvalho, et al. (2015) proposed a classifier using wide-area multilateration, accounting for GPS errors and time-of-arrival noise. Their results show that time-stamp differentials are more reliable than autocorrelations (Monteiro, Barreto, Kacem, Wijesekera, & Costa, 2015). Leonardi (2018) introduced a time-drift pattern analysis to detect masquerading ADS-B receivers. However, multilateration methods require time-stamp synchronization, which Schäfer et al. (2014) addressed using OpenSky Network data for posterior synchronization. To eliminate this dependency, Liu et al. (2019) applied Received Signal Strength Index (RSSI) and time of arrival (ToA) feature maps with an adaptive k-NN algorithm for approximate aircraft localization without coherent receivers, though precise air domain modeling remains necessary.

Velocity-Based Methods

These methods assume that spoofers do not move at the same speed as real aircraft. Velocity can be estimated using Doppler frequency spread or received signal strength variations over time. Chen et al. (2019), Ghose and Lazos (2015), and Schäfer et al. (2016) leveraged Doppler spread to detect inconsistencies between reported and actual velocity, incorporating RF channel characteristics to hinder ground-based spoofing attacks, though their approach remains unverified outside simulations. Velocity-based verification is weaker than location-based methods since adversaries can manipulate a transmitter's center frequency to mimic real transponders. Strohmeier et al. (2015) used RSSI autocorrelation and Pearson's correlation against claimed distances to assess airborne transmitters, though adversaries with precise receiver location knowledge can counteract this by adjusting ADS-B transmission power dynamically.

Open Issues

A summary of specific countermeasures to secure the ADS-B system is given in Table 8.2. Although technologies in wireless transmitter identification can become a critical foundation, we also have to point out that multi-perspective technologies must be applied to deal with various attack scenarios. In this specific domain, we can derive several conclusion through the review:

- Compared with general-purpose emitter identification, Specific Emitter Identification (SEI) in ADS-B would have to deal with the time-varying location information of aircraft as in Habler and Shabtai (2018).

TABLE 8.2
Comparison of Verification Technologies and Potential Weakness in ADS-B Systems

		Detection Capability of Malicious Activities[1]				
Methods	**Categorization**	**Ghost Aircraft**	**Transmitter Hijacking**	**Identity Spoofing**	**Weakness**	**References**
Steady radiometric fingerprinting	Device level	High	N/A	Median	• An adversary can record signal near the victim receiver and generate highly realistic signals	Chen et al. (2019)
Transient radiometric fingerprinting	Device level	High	N/A	High	• Same as above, but more difficult deal to nonlinearity of hardware	Ali et al. (2019)
Context anomaly detection	Route level	N/A	Median	N/A	• An attacker may try to mimic the pattern of original devices	Habler and Shabtai (2018)
Wide area multilateration	Aircraft level	High	Median	N/A	• The malicious may be able to block and generate fake signals to several distributed receivers	Monteiro, Barreto, Kacem, Carvalho, et al. (2015)
Coherent (MIMO) multilateration	Aircraft level	High	Median	N/A	• Same as above, but more difficult since receivers are placed within small regions	Nijsure et al. (2015)
Signal strength mapping	Aircraft level	Median	N/A	N/A	• The malicious may get the signal strength map and change signal power accordingly	Liu et al. (2019)

(Continued)

TABLE 8.2 *(Continued)*
Comparison of Verification Technologies and Potential Weakness in ADS-B Systems

		Detection Capability of Malicious Activities[1]				
Methods	**Categorization**	**Ghost Aircraft**	**Transmitter Hijacking**	**Identity Spoofing**	**Weakness**	**References**
Doppler frequency offset	Aircraft level	Median	N/A	N/A	• The malicious can change the center frequency of his transmitter to fool the signal processor	Moser et al. (2016)

[1] High: An adversary can hardly escape from the detection by transmitting specific signals. Median: In some scenarios, an adversary can escape from the detection method.

- Single type of feature is not necessarily sufficient for the non-cryptographic identification of ADS-B transponders, research in Moser et al. (2016) indicates that an adversary who has the control of multiple time-synchronized transmitters could convince multilateration-based localization systems.
- Some researchers do notice that weather and channel impairments have a potential influence on the radiometric features (Bosisio & Cadeddu, 2015; Rehman et al., 2015); however, there's no relevant research that incorporates weather effects.

Implications on Human-Machine Interface Design

Device identification is crucial in aviation security, where the accuracy and timeliness of threat detection can directly impact flight safety and ATC. Unlike general IoT security, aviation systems rely on critical real-time data transmissions, such as those from ADS-B and other avionics systems. Integrating device identification into HMIs in aviation requires balancing automation with human oversight, ensuring that security personnel and air traffic controllers can quickly verify device authenticity without being overwhelmed by excessive system complexity. While statistical learning offers lightweight and interpretable solutions for device identification, aviation systems demand higher resilience against spoofing and hijacking attempts, making deep learning and anomaly detection more suitable for detecting sophisticated cyber threats.

However, the reliance on deep learning and anomaly detection in aviation security presents challenges related to interpretability and response time. Statistical learning provides straightforward rule-based authentication, while deep learning enhances accuracy but operates as a "black box," requiring XAI for user trust. Abnormal device identification detects unknown threats but may trigger frequent false positives, adding to cognitive load for aviation security operators. To address these challenges, aviation HMIs must integrate multi-layered security interfaces that present anomaly detection results alongside historical device behavior patterns, enabling security personnel to make informed decisions. Context-aware visualizations, such as aircraft trajectory deviation mapping, ADS-B transmission fingerprinting, and automated cross-checking with ground-based radar, can provide layered security insights without overwhelming users with excessive technical details.

The high-stakes nature of aviation security also requires HMIs to prioritize response efficiency and real-time alerting. Unlike general IoT environments, where delayed anomaly detection may be acceptable, aviation systems demand immediate threat classification and mitigation strategies. Statistical learning provides rapid but limited verification, while deep learning ensures high detection accuracy but may introduce processing delays. Anomaly detection is essential for uncovering emerging threats but requires careful alert management to avoid overwhelming human operators. By designing adaptive HMIs that integrate statistical learning for baseline verification, deep learning for high-confidence classification, and anomaly detection for real-time behavioral monitoring, aviation security systems can optimize both accuracy and usability. This approach enhances human decision-making while ensuring that automated security mechanisms remain interpretable, efficient, and responsive to evolving cyber threats in aviation networks.

REFERENCES

Agadakos, I., Agadakos, N., Polakis, J., & Amer, M. R. (2019). Deep complex networks for protocol-agnostic radio frequency device fingerprinting in the wild. *arXiv preprint arXiv:1909.08703.*

Ali, A. M., Uzundurukan, E., & Kara, A. (2019). Assessment of features and classifiers for Bluetooth RF fingerprinting. *IEEE Access, 7*, 50524–50535.

Alipour, H., Al-Nashif, Y. B., Satam, P., & Hariri, S. (2015). Wireless anomaly detection based on IEEE 802.11 behavior analysis. *IEEE Transactions on Information Forensics and Security*, *10*(10), 2158–2170.

Azarmehr, M., Mehta, A., & Rashidzadeh, R. (2017). Wireless device identification using oscillator control voltage as RF fingerprint. *2017 IEEE 30th Canadian Conference on Electrical and Computer Engineering (CCECE)*, Windsor, ON, Canada, 2017, pp. 1–4. https://doi.org/10.1109/CCECE.2017.7946820.

Baldini, G., Gentile, C., Giuliani, R., & Steri, G. (2018). Comparison of techniques for radiometric identification based on deep convolutional neural networks. *Electronics Letters, 55*(2), 90–92.

Bendale, A., & Boult, T. E. (2016). Towards open set deep networks. *Proceedings of the IEEE conference on computer vision and pattern recognition*, 1563–1572.

Bergstra, J. S., Bardenet, R., Bengio, Y., & Kégl, B. (2011). Algorithms for hyper-parameter optimization. *Advances in neural information processing systems, In Proceedings of the 25th International Conference on Neural Information Processing Systems (NIPS'11). Curran Associates Inc.*, Red Hook, NY, USA, 2546–2554.

Bihl, T. J., Bauer, K. W., & Temple, M. A. (2016). Feature selection for RF fingerprinting with multiple discriminant analysis and using Zigbee device emissions. *IEEE Transactions on Information Forensics and Security, 11*(8), 1862–1874.

Bosisio, A. V., & Cadeddu, M. P. (2015). Rain detection from ground-based radiometric measurements: Validation against rain sensor observations. 2015 IEEE International Geoscience and Remote Sensing Symposium (IGARSS), 2323–2326.

Butun, I., Österberg, P., & Song, H. (2019). Security of the Internet of Things: Vulnerabilities, attacks and countermeasures. *IEEE Communications Surveys Tutorials*, 1–1. https://doi.org/10.1109/COMST.2019.2953364

Chatterjee, B., Das, D., Maity, S., & Sen, S. (2018). RF-PUF: Enhancing IoT security through authentication of wireless nodes using in-situ machine learning. *IEEE Internet of Things Journal*, *6*(1), 388–398.

Chen, S., Xie, F., Chen, Y., Song, H., & Wen, H. (2017). Identification of wireless transceiver devices using radio frequency (RF) fingerprinting based on STFT analysis to enhance authentication security. 2017 IEEE 5th International Symposium on Electromagnetic Compatibility (EMC-Beijing), 1–5.

Chen, S., Zheng, S., Yang, L., & Yang, X. (2019). Deep learning for large-scale real-world ACARS and ADS-B radio signal classification. *arXiv preprint arXiv*:1904.09425.

Chen, Y., Zhou, X. S., & Huang, T. S. (2001). One-class SVM for learning in image retrieval. *Proceedings 2001 International Conference on Image Processing (Cat. No. 01CH37205)*, *1*, 34–37.

Cobb, W. E., Garcia, E. W., Temple, M. A., Baldwin, R. O., & Kim, Y. C. (2010). Physical layer identification of embedded devices using RF-DNA fingerprinting. *2010-Milcom 2010 Military Communications Conference*, 2168–2173.

Costin, A., & Francillon, A. (2012). Ghost in the air (traffic): On insecurity of ADS-B protocol and practical attacks on ADS-B devices. Black Hat USA, pp 1–12. July 21-26, 2012, Las Vegas, NV, USA. Url: https://www.eurecom.fr/publication/3788

Cozzolino, D., & Verdoliva, L. (2020). Noiseprint: A CNN-based camera model fingerprint. *IEEE Transactions on Information Forensics and Security*. *15*, 144–159. https://doi.org/10.1109/TIFS.2019.2916364

Danev, B., Luecken, H., Capkun, S., & El Defrawy, K. (2010). Attacks on physical-layer identification. *Proceedings of the Third ACM Conference on Wireless Network Security*, 89–98.

Dartmann, G., Song, H., & Schmeink, A. (2019). *Big data analytics for cyber-physical systems: Machine learning for the Internet of Things*. Elsevier.

Deng, S., Huang, Z., Wang, X., & Huang, G. (2017). Radio frequency fingerprint extraction based on multidimension permutation entropy. *International Journal of Antennas and Propagation*, *2017*.

Ding, Z., & Fei, M. (2013). An anomaly detection approach based on isolation forest algorithm for streaming data using sliding window. *IFAC Proceedings Volumes*, *46*(20), 12–17.

Dubendorfer, C. K., Ramsey, B. W., & Temple, M. A. (2012). An RF DNA verification process for Zigbee networks. *MILCOM 2012-2012 IEEE Military Communications Conference*, 1–6.

Ghose, N., & Lazos, L. (2015). Verifying ADS-B navigation information through Doppler shift measurements. *2015 IEEE/AIAA 34th Digital Avionics Systems Conference (DASC)*, 4A2–1.

Habler, E., & Shabtai, A. (2018). Using LSTM encoder-decoder algorithm for detecting anomalous ADS-B messages. *Computers & Security*, *78*, 155–173.

Harmer, P. K., Temple, M. A., Buckner, M. A., & Farquahar, E. (2011). Using differential evolution to optimize 'learning from signals' and enhance network security. *Proceedings of the 13th annual conference on genetic and evolutionary computation*, 1811–1818.

Herder, C., Yu, M.-D., Koushanfar, F., & Devadas, S. (2014). Physical unclonable functions and applications: A tutorial. *Proceedings of the IEEE*, *102*(8), 1126–1141.

Huang, J., Lei, Y., & Liao, X. (2017). Communication transmitter individual feature extraction method based on stacked denoising autoencoders under small sample prerequisite.

2017 7th IEEE International Conference on Electronics Information and Emergency Communication (ICEIEC), 132–135.

Jafari, H., Omotere, O., Adesina, D., Wu, H.-H., & Qian, L. (2018). IoT devices fingerprinting using deep learning. *MILCOM 2018-2018 IEEE Military Communications Conference (MILCOM)*, 1–9.

Jeschke, S., Brecher, C., Song, H., & Rawat, D. (2017). *Industrial Internet of Things: Cybermanufacturing systems*. Springer.

Kim, K., Spooner, C. M., Akbar, I., & Reed, J. H. (2008). Specific emitter identification for cognitive radio with application to IEEE 802.11. *IEEE GLOBECOM 2008-2008 IEEE Global Telecommunications Conference*, 1–5.

Kirkpatrick, J., Pascanu, R., Rabinowitz, N., Veness, J., Desjardins, G., Rusu, A. A., Milan, K., Quan, J., Ramalho, T., Grabska-Barwinska, A., et al. (2017). Overcoming catastrophic forgetting in neural networks. *Proceedings of the National Academy of Sciences, 114*(13), 3521–3526.

Köse, M., Taşcioğlu, S., & Telatar, Z. (2019). RF fingerprinting of IoT devices based on transient energy spectrum. *IEEE Access, 7*, 18715–18726.

Kriegel, H.-P., Kröger, P., Schubert, E., & Zimek, A. (2009). Loop: Local outlier probabilities. *Proceedings of the 18th ACM conference on Information and knowledge management (CIKM '09). Association for Computing Machinery*, New York, NY, USA, 1649–1652. https://doi.org/10.1145/1645953.1646195.

Lei, Y.-K. (2016). Individual communication transmitter identification using correntropy-based collaborative representation. 2016 9th International Congress on Image and Signal Processing, BioMedical Engineering and Informatics (CISP-BMEI), 1194–1200.

Leonardi, M. (2018). ADS-B anomalies and intrusions detection by sensor clocks tracking. *IEEE Transactions on Aerospace and Electronic Systems. 55*(5), 2370–2381. https://doi.org/10.1109/TAES.2018.2886616.

Leonardi, M., & Di Fausto, D. (2018a). ADS-B signal signature extraction for intrusion detection in the air traffic surveillance system. *2018 26th European Signal Processing Conference (EUSIPCO)*, 2564–2568.

Leonardi, M., & Di Fausto, D. (2018b). Secondary surveillance radar transponders classification by RF fingerprinting. 2018 19th International Radar Symposium (IRS), 1–10.

Leonardi, M., Di Gregorio, L., & Di Fausto, D. (2017). Air traffic security: Aircraft classification using ADS-B message's phase-pattern. *Aerospace, 4*(4), 51.

Li, Y., Chen, L., Chen, J., Xie, F., Chen, S., & Wen, H. (2018). A low complexity feature extraction for the RF fingerprinting process. *2018 IEEE Conference on Communications and Network Security (CNS)*, 1–2.

Liu, G., Zhang, R., Wang, C., & Liu, L. (2019). Synchronization-free GPS spoofing detection with crowdsourced air traffic control data. *2019 20th IEEE International Conference on Mobile Data Management (MDM)*, 260–268.

Luo, T., & Nagarajan, S. G. (2018). Distributed anomaly detection using autoencoder neural networks in WSN for IoT. *2018 IEEE International Conference on Communications (ICC)*, 1–6.

Meidan, Y., Bohadana, M., Mathov, Y., Mirsky, Y., Shabtai, A., Breitenbacher, D., & Elovici, Y. (2018). N-BaIoT—Network-based detection of IoT botnet attacks using deep autoencoders. *IEEE Pervasive Computing, 17*(3), 12–22.

Merchant, K., Revay, S., Stantchev, G., & Nousain, B. (2018). Deep learning for RF device fingerprinting in cognitive communication networks. *IEEE Journal of Selected Topics in Signal Processing, 12*(1), 160–167.

Monteiro, M., Barreto, A., Kacem, T., Carvalho, J., Wijesekera, D., & Costa, P. (2015). Detecting malicious ADS-B broadcasts using wide area multilateration. *2015 IEEE/AIAA 34th Digital Avionics Systems Conference (DASC)*, 4A3–1.

Monteiro, M., Barreto, A., Kacem, T., Wijesekera, D., & Costa, P. (2015). Detecting malicious ADS-B transmitters using a low-bandwidth sensor network. *2015 18th International Conference on Information Fusion (Fusion)*, 1696–1701.

Morin, C., Cardoso, L., Hoydis, J., Gorce, J.-M., & Vial, T. (2019). Transmitter classification with supervised deep learning. *arXiv preprint arXiv:1905.07923.*

Moser, D., Leu, P., Lenders, V., Ranganathan, A., Ricciato, F., & Capkun, S. (2016). Investigation of multi-device location spoofing attacks on air traffic control and possible countermeasures. *Proceedings of the 22nd Annual International Conference on Mobile Computing and Networking*, 375–386.

Nijsure, Y. A., Kaddoum, G., Gagnon, G., Gagnon, F., Yuen, C., & Mahapatra, R. (2015). Adaptive air-to-ground secure communication system based on ADS-B and wide-area multilateration. *IEEE Transactions on Vehicular Technology*, *65*(5), 3150–3165.

Oyedare, T., & Park, J.-M. J. (n.d.). Estimating the required training dataset size for transmitter classification using deep learning. 2019 IEEE International Symposium on Dynamic Spectrum Access Networks (DySPAN), Newark, NJ, USA, 2019, pp. 1–10. https://doi.org/10.1109/DySPAN.2019.8935823.

Parisi, G. I., Kemker, R., Part, J. L., Kanan, C., & Wermter, S. (2019). Continual lifelong learning with neural networks: A review. *Neural Networks.* 113 (54–71). https://doi.org/10.1016/j.neunet.2019.01.012.

Parwez, M. S., Rawat, D. B., & Garuba, M. (2017). Big data analytics for user-activity analysis and user-anomaly detection in mobile wireless network. *IEEE Transactions on Industrial Informatics*, *13*(4), 2058–2065.

Pelikan, M., Goldberg, D. E., Cantú-Paz, E., et al. (1999). BOA: The Bayesian optimization algorithm. *Proceedings of the Genetic and Evolutionary Computation Conference GECCO-99*, *1*, 525–532.

Peng, L., Hu, A., Jiang, Y., Yan, Y., & Zhu, C. (2016). A differential constellation trace figure based device identification method for Zigbee nodes. *2016 8th International Conference on Wireless Communications & Signal Processing (WCSP)*, 1–6.

Peng, L., Hu, A., Zhang, J., Jiang, Y., Yu, J., & Yan, Y. (2018). Design of a hybrid RF fingerprint extraction and device classification scheme. *IEEE Internet of Things Journal*, *6*(1), 349–360.

Peng, L., Zhang, J., Liu, M., & Hu, A. (2019). Deep learning based RF fingerprint identification using differential constellation trace figure. *IEEE Transactions on Vehicular Technology*. *69*(1), 1091–1095. https://doi.org/10.1109/TVT.2019.2950670.

Polak, A. C., & Goeckel, D. L. (2015a). Identification of wireless devices of users who actively fake their RF fingerprints with artificial data distortion. *IEEE Transactions on Wireless Communications*, *14*(11), 5889–5899.

Polak, A. C., & Goeckel, D. L. (2015b). Wireless device identification based on RF oscillator imperfections. *IEEE Transactions on Information Forensics and Security*, *10*(12), 2492–2501.

Rajendran, S., Meert, W., Lenders, V., & Pollin, S. (2018). SAIFE: Unsupervised wireless spectrum anomaly detection with interpretable features. 2018 IEEE International Symposium on Dynamic Spectrum Access Networks (DySPAN), 1–9.

Ramsey, B. W., Stubbs, T. D., Mullins, B. E., Temple, M. A., & Buckner, M. A. (2015). Wireless infrastructure protection using low-cost radio frequency fingerprinting receivers. *International Journal of Critical Infrastructure Protection*, *8*, 27–39.

Rehman, S. U., Sowerby, K. W., Alam, S., Ardekani, I. T., & Komosny, D. (2015). Effect of channel impairments on radiometric fingerprinting. 2015 IEEE International Symposium on Signal Processing and Information Technology (ISSPIT), 415–420.

Reising, D. R., Temple, M. A., & Jackson, J. A. (2015). Authorized and rogue device discrimination using dimensionally reduced RF-DNA fingerprints. *IEEE Transactions on Information Forensics and Security*, *10*(6), 1180–1192.

Riyaz, S., Sankhe, K., Ioannidis, S., & Chowdhury, K. (2018). Deep learning convolutional neural networks for radio identification. *IEEE Communications Magazine*, *56*(9), 146–152.

Roy, D., Mukherjee, T., Chatterjee, M., Blasch, E., & Pasiliao, E. (2019). RFAL: Adversarial learning for RF transmitter identification and classification. *IEEE Transactions on Cognitive Communications and Networking*, *6*(2), 783–801.

Roy, D., Mukherjee, T., Chatterjee, M., & Pasiliao, E. (2019). Detection of rogue RF transmitters using generative adversarial nets. In *2019 IEEE wireless communications and networking conference (WCNC)*. Marrakesh, Morocco, 2019, pp. 1–7. https://doi.org/10.1109/WCNC.2019.8885548.

Satija, U., Trivedi, N., Biswal, G., & Ramkumar, B. (2018). Specific emitter identification based on variational mode decomposition and spectral features in single hop and relaying scenarios. *IEEE Transactions on Information Forensics and Security*, *14*(3), 581–591.

Schäfer, M., Leu, P., Lenders, V., & Schmitt, J. (2016). Secure motion verification using the Doppler effect. *Proceedings of the 9th ACM Conference on Security & Privacy in Wireless and Mobile Networks*, 135–145.

Schäfer, M., Strohmeier, M., Lenders, V., Martinovic, I., & Wilhelm, M. (2014). Bringing up OpenSky: A large-scale ADS-B sensor network for research. *Proceedings of the 13th international symposium on Information processing in sensor networks*, 83–94.

Schäfer, M., Strohmeicr, M., Smith, M., Fuchs, M., Lenders, V., & Martinovic, I. (2018). OpenSky report 2018: Assessing the integrity of crowdsourced mode s and ADS-B data. *2018 IEEE/AIAA 37th Digital Avionics Systems Conference (DASC)*, 1–9.

Scheirer, W. J., de Rezende Rocha, A., Sapkota, A., & Boult, T. E. (2012). Toward open set recognition. *IEEE Transactions on Pattern Analysis and Machine Intelligence*, *35*(7), 1757–1772.

Sehatbakhsh, N., Alam, M., Nazari, A., Zajic, A., & Prvulovic, M. (2018). Syndrome: Spectral analysis for anomaly detection on medical IoT and embedded devices. *2018 IEEE international symposium on hardware oriented security and trust (HOST)*, 1–8.

Shi, Y., Davaslioglu, K., Sagduyu, Y. E., Headley, W. C., Fowler, M., & Green, G. (2019). Deep learning for RF signal classification in unknown and dynamic spectrum environments. *2019 IEEE International Symposium on Dynamic Spectrum Access Networks (DySPAN)*, 1–10.

Shi, Z., Lin, X., Zhao, C., & Shi, M. (2015). Multifractal slope feature based wireless devices identification. *2015 10th International Conference on Computer Science & Education (ICCSE)*, 590–595.

Shin, H., Lee, J. K., Kim, J., & Kim, J. (2017). Continual learning with deep generative replay. In I. Guyon, U. Von Luxburg, S. Bengio, H. Wallach, R. Fergus, & S. Vishwanathan, & R. Garnett (Eds.), *Advances in Neural Information Processing Systems* (vol. 30). Curran Associates, Inc., 2990–2999.

Song, H., Fink, G. A., & Jeschke, S. (2017). *Security and privacy in cyber-physical systems: Foundations, principles and applications*. Wiley-IEEE Press.

Sridharan, R., Maiti, R. R., & Tippenhauer, N. O. (2018). WADAC: Privacy-preserving anomaly detection and attack classification on wireless traffic. *Proceedings of the 11th ACM Conference on Security & Privacy in Wireless and Mobile Networks*, 51–62.

Strohmeier, M. (2017). Large-scale analysis of aircraft transponder data. *IEEE Aerospace and Electronic Systems Magazine*, *32*(1), 42–44.

Strohmeier, M., Lenders, V., & Martinovic, I. (2014). On the security of the Automatic Dependent Surveillance-Broadcast protocol. *IEEE Communications Surveys & Tutorials*, *17*(2), 1066–1087.

Strohmeier, M., Lenders, V., & Martinovic, I. (2015). Intrusion detection for airborne communication using PHY-layer information. *International Conference on Detection of Intrusions and Malware, and Vulnerability Assessment*, 67–77.

Strohmeier, M., & Martinovic, I. (2015). On passive data link layer fingerprinting of aircraft transponders. *Proceedings of the First ACM Workshop on Cyber-Physical Systems-Security and/or PrivaCy*, 1–9.

Sun, D., Li, Y., & Xu, Y. (2016). Specific emitter identification based on normalized frequency spectrum. *2016 2nd IEEE International Conference on Computer and Communications (ICCC)*, 1875–1879.

Sun, J. (2017, May). The 1090 Megahertz Riddle: An open-access book about decoding mode-s and ADS-B data. TU Delft OPEN Publishing,, https://doi.org/10.34641/mg.11.

Sun, M., Zhang, L., Bao, J., & Yan, Y. (2017). RF fingerprint extraction for GNSS anti-spoofing using axial integrated Wigner bispectrum. *Journal of Information Security and Applications*, *35*, 51–54.

Tan, C., Sun, F., Kong, T., Zhang, W., Yang, C., & Liu, C. (2018). A survey on deep transfer learning. *International Conference on Artificial Neural Networks*, 270–279.

Tandiya, N., Jauhar, A., Marojevic, V., & Reed, J. H. (2018). Deep predictive coding neural network for RF anomaly detection in wireless networks. *2018 IEEE International Conference on Communications Workshops (ICC Workshops)*, 1–6.

Thing, V. L. (2017). IEEE 802.11 network anomaly detection and attack classification: A deep learning approach. *2017 IEEE Wireless Communications and Networking Conference (WCNC)*, 1–6.

Truong, A., Walters, A., Goodsitt, J., Hines, K., Bruss, C. B., & Farivar, R. (2019, November). Towards automated machine learning: Evaluation and comparison of AutoML approaches and tools. In 2019 IEEE 31st international conference on tools with artificial intelligence (ICTAI) pp. 1471–1479). IEEE.

Tu, Y., Zhang, Z., Li, Y., Wang, C., & Xiao, Y. (2019). Research on the Internet of Things device recognition based on RF-fingerprinting. *IEEE Access*, *7*, 37426–37431.

Wang, K.-H., Chen, C.-M., Fang, W., & Wu, T.-Y. (2018). On the security of a new ultra-lightweight authentication protocol in IoT environment for RFID tags. *The Journal of Supercomputing*, *74*(1), 65–70.

Wang, X., Duan, J., Wang, C., Cui, G., & Wang, W. (2017). A radio frequency fingerprinting identification method based on energy entropy and color moments of the bispectrum. *2017 9th International Conference on Advanced Infocomm Technology (ICAIT)*, 150–154.

Wei, M., & Kim, K. (2012). Intrusion detection scheme using traffic prediction for wireless industrial networks. *Journal of Communications and Networks*, *14*(3), 310–318.

Wong, L. J., Headley, W. C., Andrews, S., Gerdes, R. M., & Michaels, A. J. (2018). Clustering learned CNN features from raw I/Q data for emitter identification. *MILCOM 2018-2018 IEEE Military Communications Conference (MILCOM)*, 26–33.

Wu, X., Shi, Y., Meng, W., Ma, X., & Fang, N. (2019). Specific emitter identification for satellite communication using probabilistic neural networks. *International Journal of Satellite Communications and Networking*, *37*(3), 283–291.

Ying, X., Mazer, J., Bernieri, G., Conti, M., Bushnell, L., & Poovendran, R. (2019). Detecting ADS-B spoofing attacks using deep neural networks. *arXiv preprint arXiv:1904.09969.*

Yoon, J., Yang, E., Lee, J., & Hwang, S. J. (2017). Lifelong learning with dynamically expandable networks. *arXiv preprint arXiv:1708.01547.*

Young, S. R., Rose, D. C., Karnowski, T. P., Lim, S.-H., & Patton, R. M. (2015). Optimizing deep learning hyper-parameters through an evolutionary algorithm. *Proceedings of the Workshop on Machine Learning in High-Performance Computing Environments*, 1–5.

Youssef, K., Bouchard, L., Haigh, K., Silovsky, J., Thapa, B., & Vander Valk, C. (2018). Machine learning approach to RF transmitter identification. *IEEE Journal of Radio Frequency Identification*, *2*(4), 197–205.

Yu, J., Hu, A., Li, G., & Peng, L. (2019). A robust RF fingerprinting approach using multi-sampling convolutional neural network. *IEEE Internet of Things Journal*, *6*(4), 6786–6799. https://doi.org/10.1109/JIOT.2019.2911347.

Yu, J., Hu, A., Zhou, F., Xing, Y., Yu, Y., Li, G., & Peng, L. (2019). Radio frequency fingerprint identification based on denoising autoencoders. *arXiv preprint arXiv:1907.08809.*

Yuan, Y., Huang, Z., Wu, H., & Wang, X. (2014). Specific emitter identification based on Hilbert–Huang transform-based time–frequency–energy distribution features. *IET Communications*, *8*(13), 2404–2412.

Zhang, Z., Li, Y., Wang, C., Wang, M., Tu, Y., & Wang, J. (2018). An ensemble learning method for wireless multimedia device identification. *Security and Communication Networks*, *2018*.

Zhuang, Z., Ji, X., Zhang, T., Zhang, J., Xu, W., Li, Z., & Liu, Y. (2018). FBSleuth: Fake base station forensics via radio frequency fingerprinting. *Proceedings of the 2018 on Asia Conference on Computer and Communications Security*, 261–272.

9 Cybersecurity and Ground Transportation

Xiaojian Jin

ACKNOWLEDGEMENTS

I would like to extend my sincere gratitude to Crystal for inviting me to contribute to this textbook. Her encouragement and support made this chapter possible. Additionally, much of the content in this section originates from my previous unpublished technical report, and I appreciate the opportunity to refine and present these ideas in a broader academic context.

INTRODUCTION

The integration of cybersecurity in ground transportation has become increasingly critical as commercial motor vehicle (CMV) fleets adopt automated driving systems (ADS). With the transition from conventional trucks to mixed fleets, including both ADS-equipped and traditional vehicles, the transportation industry faces new cybersecurity challenges. While automation brings efficiency and safety improvements, it also introduces vulnerabilities that cyber threats can exploit.

Cybersecurity in ground transportation is not just a technical issue; it involves human factors, policy considerations, and operational complexities. The safe and secure deployment of ADS technology depends on a comprehensive approach that addresses potential cyber risks, protects critical vehicle systems, and ensures that fleet operators and drivers are equipped to handle security threats.

This chapter explores the cybersecurity landscape of modern CMV fleets, highlighting the vulnerabilities, best practices, and future challenges associated with cybersecurity in mixed fleets. It provides a balanced discussion of technical security measures, operational risk management, and human factors considerations to help fleet operators, policymakers, and ADS developers mitigate cybersecurity threats effectively.

By examining key cybersecurity risks – such as data security, wireless intrusions, unauthorized access, and supply chain vulnerabilities, this chapter aims to provide practical guidance on securing ground transportation systems. Additionally, it will address specialized use cases such as teleoperation, cooperative perception, and vehicle platooning, where cybersecurity plays a vital role in ensuring safe operations.

As the transportation industry continues to evolve, cybersecurity must be treated as a fundamental pillar of safety and operational integrity. A proactive approach, including continuous monitoring, cybersecurity training, and adherence to best practices, will be essential to maintaining the security and resilience of modern transportation networks.

DOI: 10.1201/9781003413707-10

BACKGROUND ON CYBERSECURITY IN TRANSPORTATION

Definition of Cybersecurity and Its Relevance in CMVs

Cybersecurity in ground transportation refers to the protection of vehicles, networks, and data systems from unauthorized access, cyberattacks, and data breaches. As modern CMVs incorporate connected and automated technologies, they rely on software, wireless communication, and data exchanges to operate efficiently. While these advancements improve fleet management, safety, and logistics, they also introduce new cybersecurity risks that require proactive mitigation strategies.

The Cybersecurity & Infrastructure Security Agency (CISA) defines cybersecurity as "the art of protecting networks, devices, and data from unauthorized access or criminal use and ensuring the confidentiality, integrity, and availability of information." In the context of transportation, cybersecurity extends beyond traditional IT security and must account for vehicle automation, wireless communication, remote fleet management, and data transfer between multiple stakeholders.[1]

In 2015, two security researchers exposed the security vulnerabilities in automobiles by hacking into cars remotely, overtaking the cars' various controls from the radio volume to the brakes,[2] and leading to 1.4 million vehicles being recalled from the car manufacturer. In 2020, a Tennessee-based trucking and logistic company was targeted by a new and relatively unknown group of ransomware operators called "Hades" in December, which caused damage to the fleet's reputation and disrupted operation.[3] Meanwhile, the information shared among ADSs and between the ADSs and infrastructure can trigger safety-critical functions on the vehicles. Recent studies have revealed concerns of major risks associated with connected and automated vehicles as the technology advances; these risks include not only the risk of traditional cyberattacks on the information and running of the vehicle, but also a new breed of attacks around such things as ransomware, Internet of Things (IoT) attacks, and vehicle theft.[4]

While cybersecurity focuses on digital threats, security in ground transportation is a broader concept that includes both cyber and physical safety. This chapter will use these terms interchangeably, acknowledging that cybersecurity measures ultimately contribute to the overall safety of transportation systems.

Cybersecurity vs. General Security in Transportation Systems

General security in transportation has traditionally focused on preventing physical risks that could impact vehicles, cargo, and infrastructure. These measures include securing fleet depots, warehouses, and truck stops to prevent unauthorized access, implementing GPS tracking and security seals to deter cargo theft, and ensuring the safety of drivers and personnel through proper training, access control, and emergency response procedures. Additionally, regulatory compliance plays a crucial role in maintaining transportation security, requiring adherence to federal and state laws that govern physical threats such as break-ins, theft, and unauthorized use of CMVs. These security efforts primarily address tangible risks, ensuring that transportation assets and personnel remain protected from direct harm.

However, as transportation systems become increasingly automated and interconnected, cybersecurity has emerged as a critical area of concern. Unlike general security, which primarily deals with physical threats, cybersecurity focuses on protecting data confidentiality, maintaining system integrity, and ensuring operational continuity. Cyber threats can originate from various sources, including remote cyberattacks that target network vulnerabilities in ADS-equipped CMVs, data breaches that expose fleet management and logistics information, and malware or ransomware attacks that encrypt fleet data and disrupt operations. Additionally, GPS spoofing and manipulation pose significant risks by allowing malicious actors to alter location data, misleading fleet operations and facilitating cargo theft.

The interconnection between cybersecurity and general security is becoming more pronounced as digital and physical threats increasingly overlap. A cyberattack on fleet management systems could have tangible consequences, such as disrupting supply chain logistics, preventing ADS-equipped trucks from executing driving commands, or overriding security tracking mechanisms to facilitate cargo theft. Likewise, physical security breaches can lead to cyber vulnerabilities, as unauthorized personnel gaining access to vehicle control units may tamper with onboard diagnostic (OBD) systems or introduce malware through external devices plugged into CMV telematics systems. Given these interdependencies, cybersecurity must be fully integrated into general transportation security strategies. Fleet operators, ADS developers, and policymakers must collaborate to establish comprehensive security measures that address both physical and cyber risks, ensuring a robust and holistic approach to transportation security.

ROLE OF KEY STAKEHOLDERS

Ensuring cybersecurity in ground transportation is a shared responsibility among multiple stakeholders, each playing a crucial role in preventing cyber threats.

Fleet Operators and Owners

Fleet operators and owners are responsible for implementing cybersecurity best practices in their daily operations. They must:

- Maintain secure network infrastructure for fleet management.
- Train personnel on cybersecurity awareness.
- Ensure compliance with cybersecurity policies and regulations.

ADS Developers and Technology Providers

Companies developing ADS technology and vehicle software must incorporate cybersecurity into the design and testing of automated trucks. This includes:

- Conducting risk assessments and penetration testing.
- Implementing secure software update mechanisms.
- Designing robust access control and encryption protocols.

Policymakers and Regulators

Government agencies such as Federal Motor Carrier Safety Administration (FMCSA), National Highway Traffic Safety Administration (NHTSA), and CISA set cybersecurity guidelines for commercial vehicles and automated driving technologies. Their role includes:

- Establishing cybersecurity policies and compliance standards.
- Conducting research and issuing recommendations.
- Collaborating with industry stakeholders to address emerging threats.

Drivers and Operational Personnel

Although cybersecurity is often viewed as a technical issue, human factors play a significant role. Drivers and fleet personnel must be trained to:

- Recognize cybersecurity threats such as phishing attempts or unauthorized network access.
- Follow cybersecurity protocols for logging into fleet management systems.
- Respond to security incidents, including reporting suspicious activity.

CYBERSECURITY CHALLENGES IN MIXED FLEETS

As CMV fleets transition to mixed fleet operations – integrating both traditional and ADS-equipped trucks, new cybersecurity challenges emerge. Unlike conventional fleets, mixed fleets rely heavily on data-driven operations, vehicle connectivity, and automated decision-making, increasing their exposure to cyber threats.

The complexity of coordinating mixed fleets creates unique vulnerabilities related to software integrity, communication networks, data security, and remote access points. Fleet operators must address these challenges proactively to ensure safe and secure transportation operations. Figure 9.1 below is a fleet Threat Examples Severity Matrix that maps some examples of threats with their severity, this does not conclude all threats, these are only a few selected common examples to represent their impact levels under their threat categories. The content after this matrix further discusses threats in details.

SOFTWARE AND COMMUNICATION VULNERABILITIES

Software and communication networks form the backbone of modern transportation systems, facilitating fleet management, vehicle-to-vehicle (V2V) communication, and real-time ADS functionality. However, as connectivity increases, so do the attack surfaces for cybercriminals.

In mixed fleets, cybersecurity vulnerabilities often stem from:

- Interconnected vehicle systems that expose control functions to cyber threats.
- Weak encryption or authentication measures that allow unauthorized access.

Severity \ Threats	Vehicle Control & Safety	Data Breach & Privacy	Supply Chain & Third-Party Risks	Infrastructure & Operational Disruption	Insider Threats & Human Factors
Low Impact	Unauthorized access to vehicle systems	Unauthorized access to fleet data	Inventory error	DDoS* attacks on operations	Poor cyber hygiene
Medium Impact	Spoofing of sensor data	Sensitive customer information theft	Insecure IoT* devices	GPS spoofing	Unintentional data leaks
High Impact	Manipulation of automated systems	Data tampering	Compromised software updates	Critical infrastructure tampering	Insider attacks
Critical Impact	Remote vehicle takeover	Ransomware attacks	Fraudulent parts	Malware attacks	Social engineering exploits

FIGURE 9.1 Threat examples severity matrix.

- Software bugs or outdated firmware that create exploitable security gaps.
- Wireless threats such as GPS spoofing, IoT hacking, and network intrusions.

The following subsections explore two major categories of software and communication vulnerabilities: interconnected systems in ADS-equipped CMVs and wireless cyber threats such as GPS spoofing and IoT vulnerabilities.

Risks of Interconnected Systems in ADS-Equipped CMVs

Unlike conventional trucks, ADS-equipped CMVs rely on a complex network of sensors, software algorithms, and data exchanges to make autonomous driving decisions. These interconnected systems introduce multiple cybersecurity risks, as vulnerabilities in one component can compromise the entire vehicle or even an entire fleet.

One of the key risks in mixed fleets is the increased attack surface due to continuous data exchanges between perception systems such as Light Detection and Ranging (LiDAR), radar, and cameras, as well as control algorithms and external networks. A single vulnerability, such as an exposed communication link or outdated firmware, can be exploited to gain unauthorized access, allowing a cyberattack on one truck to propagate across interconnected systems. Additionally, software integrity risks pose a significant threat, as ADS-equipped vehicles require frequent updates to enhance functionality and patch security flaws. Poorly secured update mechanisms could allow attackers to inject malware, altering vehicle behavior and potentially overriding critical controls.

Remote hacking and unauthorized access further exacerbate cybersecurity concerns. Cloud-based fleet management platforms enable real-time data sharing, but without robust encryption and access control, hackers could exploit these systems to disable vehicle operations, manipulate driving commands such as steering and braking, or steal sensitive fleet data. For instance, a remote command injection attack could override ADS safety mechanisms, creating hazardous driving conditions. Similarly, data manipulation risks can mislead ADS decision-making, as compromised sensor inputs – such as falsified camera feeds or radar signals – could deceive the system into making incorrect driving decisions. A sensor spoofing attack, for example, could trick an ADS into detecting a nonexistent obstacle, triggering an unnecessary emergency stop.

To mitigate these threats, fleet operators and ADS developers must implement rigorous cybersecurity measures. End-to-end encryption should be applied to vehicle-to-cloud communication to protect data integrity. Multi-factor authentication (MFA) is essential for securing remote system access, while regular audits and secure firmware updates help address vulnerabilities before they can be exploited. Additionally, deploying intrusion detection systems (IDS) can enhance real-time monitoring, identifying network anomalies that may indicate cyber threats. By adopting these strategies, transportation stakeholders can reinforce the security of ADS-equipped CMVs and reduce the risks posed by cyberattacks.

GPS Spoofing, Wireless Intrusions, and IoT Vulnerabilities

The wireless nature of modern CMV fleets introduces additional cybersecurity risks, as cybercriminals can exploit vulnerabilities in GPS signals, wireless networks, and IoT devices integrated into fleet operations. These risks pose serious threats to navigation, communication, and vehicle control, making cybersecurity a critical concern in automated transportation systems.

GPS spoofing attacks are one of the most significant threats, as ADS-equipped CMVs rely on GPS for navigation, geofencing, and dispatch tracking. Cybercriminals can manipulate GPS signals to alter location data, enabling cargo theft, misleading vehicle tracking systems, and causing unintended route diversions. For instance, in a cargo theft operation, attackers may use GPS spoofing to make a fleet's tracking system believe a truck is following its assigned route while it is actually being rerouted to an unauthorized location. To mitigate these risks, fleet operators must implement GPS authentication techniques such as multi-source positioning verification, deploy anti-spoofing algorithms to detect false signals, and cross-reference GPS data with other navigation systems, including inertial measurement units and LiDAR-based localization.

Wireless network intrusions also present a serious risk, as fleet-wide communication relies on cellular networks, Wi-Fi, and satellite links. Attackers can intercept or manipulate network traffic through man-in-the-middle attacks, where they intercept vehicle data before it reaches the fleet control center, denial-of-service (DoS) attacks that overload communication channels to disrupt operations, and unauthorized remote access by exploiting weak authentication mechanisms. A hacker, for example, could launch a DoS attack on a fleet's telematics system, preventing ADS-equipped trucks from receiving critical operational updates. To counter these threats, fleet operators should encrypt all wireless communications using transport layer security and advanced encryption standard, deploy virtual private networks (VPNs) to create secure communication tunnels, and implement real-time anomaly detection systems to identify unusual network activity before it escalates.

IoT device exploitation is another major concern, as modern CMVs incorporate IoT sensors for cargo monitoring, fuel efficiency tracking, and predictive maintenance. These devices are often connected to public or poorly secured networks, making them easy targets for cyberattacks. A compromised IoT sensor could allow attackers to manipulate vehicle diagnostics or disable critical functions. For example, vulnerabilities in IoT-based tire pressure monitoring systems could be exploited to transmit false readings, triggering unnecessary maintenance alerts and delaying fleet operations. To reduce these risks, all IoT devices should be secured with strong authentication and encrypted data transmission, isolated from critical vehicle control networks to prevent unauthorized access, and regularly updated with firmware patches to close known security vulnerabilities.

By addressing these wireless security threats, fleet operators and ADS developers can strengthen the resilience of CMV fleets against cyberattacks, ensuring the integrity and reliability of automated transportation systems.

Data Security Risks

As CMV fleets become increasingly connected, data security has become a critical concern. Modern fleets generate, process, and transmit vast amounts of data related to vehicle diagnostics, driver behavior, operational status, and logistics. Protecting this data from cyber threats is essential to maintaining operational integrity and preventing unauthorized access. Mixed fleets, which include both conventional and ADS-equipped CMVs, introduce additional risks due to their reliance on cloud-based fleet management, wireless communication, and third-party data services.

Data Transfer Issues in Fleet Operations

Data is a fundamental component of modern fleet operations, enabling real-time tracking, automated decision-making, and seamless V2V and vehicle-to-infrastructure (V2I) communication. However, as data moves between vehicles, fleet management centers, and external systems, it becomes vulnerable to cyber threats that can compromise security, integrity, and operational efficiency.

One of the primary risks lies in data storage and processing. Fleet management platforms house vast amounts of sensitive data, including vehicle performance logs, driver records, and operational schedules. Poorly secured databases can become targets for cyberattacks such as data breaches, ransomware incidents, and insider threats. Additionally, cloud-based data processing requires continuous encryption and authentication measures to prevent unauthorized access. If security protocols are weak or outdated, attackers could intercept, manipulate, or even erase critical data, potentially disrupting fleet operations.

Data transmission vulnerabilities further heighten security risks, as CMVs rely on multiple wireless communication channels – including cellular networks, satellite links, and Wi-Fi – to transmit operational data. These communication channels are susceptible to cyberattacks such as unauthorized interception, signal jamming, and man-in-the-middle attacks. Automated trucks, in particular, use telematics systems to exchange real-time updates with fleet control centers, but without robust encryption and authentication measures, this data can be intercepted, modified, or blocked. For example, a cybercriminal could alter vehicle-to-cloud communication to manipulate trip routing data, causing fleet-wide disruptions. Similarly, attackers could exploit weak encryption in a fleet's telematics system to inject false diagnostics data, leading to unnecessary maintenance and vehicle downtime.

To mitigate these threats, fleet operators must implement comprehensive data protection strategies. End-to-end encryption should be used to secure data during both storage and transmission, while cloud platforms should incorporate strong authentication protocols and role-based access controls (RBAC) to limit unauthorized access. Regular audits of telematics and fleet management systems are essential for identifying and addressing security vulnerabilities before they can be exploited. Additionally, deploying real-time anomaly detection systems can help monitor unusual data transfer patterns, enabling early detection and response to potential cyber threats. By prioritizing these security measures, fleet operators can safeguard their data infrastructure and ensure the reliability of modern CMV operations.

Risks of Unauthorized Access, Data Corruption, and Transmission Errors

Beyond data transfer vulnerabilities, fleet operators must also address risks related to unauthorized access, data corruption, and transmission errors, all of which can compromise the integrity and reliability of fleet management systems.

Unauthorized access to fleet data is a significant concern, as cybercriminals can exploit weak login credentials, phishing attacks, or stolen authentication tokens to infiltrate fleet management platforms. Once inside, attackers may manipulate operations by rerouting shipments, disabling security settings, or causing other disruptions. Insider threats further exacerbate these risks, as malicious employees or compromised third-party service providers can leak sensitive data or intentionally disrupt system functionality.

Data corruption presents another major challenge, as unauthorized alterations to stored data can lead to inaccurate fleet analytics, incorrect route planning, and malfunctions in automated decision-making processes. Cyberattacks such as malware infections and ransomware further increase these risks by encrypting fleet data and rendering it inaccessible until a ransom is paid. The consequences of such attacks can be severe, affecting fleet efficiency and safety.

Transmission errors and data integrity issues also pose significant threats, particularly in mixed fleets where ADS-equipped trucks rely on continuous data streams for safe operation. Poor network configurations, software bugs, or cyber interference can disrupt these transmissions, leading to operational inefficiencies or unsafe driving conditions. Additionally, unverified data sources or external inputs can introduce misleading or incorrect information into fleet management systems, affecting decision-making and operational planning.

For example, an attacker who gains unauthorized access to a fleet's dispatch system could manipulate cargo schedules, causing widespread delays. Ransomware attacks targeting vehicle diagnostics data could prevent operators from accessing essential maintenance reports, leading to potential vehicle failures. Meanwhile, poorly secured telematics systems may experience transmission errors that result in inaccurate location tracking, affecting fleet coordination and route efficiency.

To mitigate these threats, fleet operators should implement MFA and enforce strict access controls across all fleet data systems to prevent unauthorized access. Regular data backups to secure, offline storage can help mitigate ransomware risks by ensuring recovery options remain available. Additionally, data integrity verification techniques such as cryptographic hashing should be employed to detect unauthorized modifications. Network monitoring solutions should also be in place to identify unauthorized access attempts in real time, allowing fleet operators to respond swiftly to potential cyber threats. By integrating these security measures, fleet operators can enhance data protection and maintain the operational integrity of their systems.

Physical Access Risks

While cybersecurity often focuses on digital threats, physical access risks also pose significant challenges for fleet security. Unauthorized physical access to CMVs and their associated infrastructure can compromise both safety and cybersecurity. Attackers who gain direct access to a vehicle's hardware, data ports, or electronic control units (ECUs) may bypass digital security measures and introduce malicious software, disable safety features, or steal sensitive fleet data.

Physical security threats are especially concerning mixed fleets, which integrate both traditional and ADS-equipped trucks. Autonomous and connected vehicles rely on sensors, electronic components, and external computing units to operate, making them particularly vulnerable to hardware-based attacks. Ensuring strong physical security measures is essential to protecting fleet assets from cyber-physical intrusions.

Potential for Hardware Tampering

Hardware tampering involves unauthorized modifications or manipulations of a vehicle's physical components, posing significant risks to CMVs. Attackers may target these vehicles by accessing OBD ports, altering telematics systems, or interfering with sensor inputs. Such attacks can lead to severe consequences, including loss of vehicle control, data breaches, and operational disruptions.

One major form of hardware tampering involves direct access to a truck's ECUs, telematics modules, or networked sensors. Attackers with physical access can alter system configurations, disable safety features, or inject malicious firmware. Another method is the installation of rogue devices, such as GPS jammers, unauthorized Wi-Fi hotspots, or skimming tools, which can be used to intercept vehicle data or disrupt communication. Tampering with cameras and sensors is also a significant threat, as ADS-equipped trucks rely on LiDAR, radar, and camera systems to perceive their surroundings. Physical interference, such as blocking or repositioning these sensors, can impair the vehicle's ability to operate safely. Additionally, unauthorized diagnostic access through OBD-II or manufacturer-specific interfaces may allow attackers to extract sensitive fleet data or reprogram vehicle control settings.

For instance, an attacker at a truck stop could gain access to a vehicle's OBD port and install a malicious device that intercepts fleet data in real time. A rogue employee might manipulate a truck's GPS module to alter route tracking and conceal unauthorized detours. In another scenario, criminals could disable forward-facing cameras and sensors on an autonomous truck, creating blind spots that increase the risk of undetected collisions.

To mitigate these risks, fleet operators must implement strict security measures. Restricting access to diagnostic ports and vehicle control interfaces through physical locks or encrypted authentication mechanisms can prevent unauthorized access. Installing tamper-evident seals and security enclosures around critical electronic components can help detect and deter tampering attempts. Fleet-wide monitoring systems should be deployed to identify unauthorized hardware modifications or suspicious network connections. Additionally, drivers and fleet personnel should receive training to recognize signs of tampering and report any suspicious activity promptly. By adopting these strategies, fleet operators can enhance the security of CMVs and reduce the risks associated with hardware tampering.

Security Concerns at Rest Stops, Fuel Stations, and Warehouses

Fleet vehicles frequently stop at rest areas, fueling stations, and distribution centers during transit, exposing them to heightened security risks. These locations present opportunities for unauthorized access, equipment theft, and cyber-physical attacks. Criminals and malicious actors may exploit these vulnerabilities to tamper with vehicles, install unauthorized tracking devices, or infiltrate fleet management systems.

Rest stops and parking areas pose significant risks, as long-haul trucks are often left unattended for extended periods. Attackers can take advantage of this by installing skimming devices, accessing telematics modules, or interfering with cargo monitoring systems. Fuel stations also present a major security concern, as CMVs rely on them for refueling, making them prime targets for card skimming, Radio Frequency Identification (RFID) cloning, and cyber-enabled fuel fraud. Unsecured Wi-Fi networks at fuel stations further increase the risk of cybercriminals intercepting fleet payment transactions. Warehouses and loading docks, critical points in the supply chain, are another area of concern, as unauthorized personnel may exploit security gaps to gain access to fleet management systems or manipulate cargo tracking data.

For example, a cybercriminal at a truck stop might connect a rogue device to an unattended CMV, gaining remote access to its fleet management system. Attackers could also exploit an unprotected warehouse network to inject false cargo tracking information into a fleet's logistics system, leading to shipment mismanagement or theft. Similarly, criminals installing RFID skimming devices at fuel pumps could steal fleet payment credentials, resulting in fraudulent transactions and financial losses.

To mitigate these risks, fleet operators should prioritize secure parking locations equipped with surveillance cameras and restricted access controls. Enforcing strict vehicle access policies can prevent unauthorized personnel from approaching fleet trucks at rest stops, reducing the risk of physical and cyber threats. Secure, encrypted fleet payment systems should be implemented, and drivers should avoid using unsecured public Wi-Fi networks for transactions. Additionally, regular vehicle inspections should be conducted to detect signs of physical tampering or unauthorized device installations. By implementing these measures, fleet operators can strengthen security at common stop locations and protect their vehicles from cyber and physical threats.

SUPPLY CHAIN SECURITY RISKS

Supply chain security is a critical aspect of cybersecurity in mixed fleets. Modern CMVs, especially those equipped with ADS, depend on a vast network of third-party suppliers for hardware, software, and communication systems. The complexity of these supply chains introduces potential vulnerabilities that cybercriminals can exploit to compromise vehicle security.

Attacks on the supply chain may occur at any stage, from the manufacturing of vehicle components to software updates and maintenance services. If a single element within the supply chain is compromised, it can have far-reaching consequences for fleet safety, data integrity, and operational reliability.

Vulnerabilities in Third-Party Software and Hardware Components

Fleet operations depend on a network of external vendors to supply critical vehicle components, including sensors, communication modules, and fleet management software. However, these third-party components, if not properly secured, can introduce cybersecurity risks that affect the entire fleet, making supply chain security a crucial consideration for fleet operators.

One of the primary concerns is the security of third-party software. Many CMVs rely on over-the-air updates to receive security patches and software enhancements, but if an attacker manages to inject malicious code into an update, they could exploit system vulnerabilities or take control of fleet management functions. Additionally, third-party fleet management applications often use application programming interfaces (APIs) to communicate with vehicles and logistics platforms. If these APIs are weak or improperly secured, attackers could gain unauthorized access to fleet data or even manipulate vehicle controls. A lack of software integrity verification further compounds the problem, as fleet operators that do not have mechanisms to validate the authenticity of third-party software risk deploying compromised or altered code that could undermine fleet security.

Beyond software risks, third-party hardware components also present security challenges. Many vehicle manufacturers source ECUs from external vendors, and if an ECU is compromised, it could introduce malware or provide a backdoor for remote attackers to manipulate vehicle operations. Counterfeit or substandard hardware components also pose a serious threat, as they may contain security vulnerabilities that increase the risk of cyber intrusions. These components can enter the supply chain without fleet operators' knowledge, leading to potential failures or exploitations. Additionally, if third-party manufacturers do not implement strong encryption or secure boot mechanisms, hardware components may be left vulnerable to tampering or unauthorized reprogramming.

For example, if a third-party telematics provider experiences a data breach, attackers could gain access to fleet location data and reroute shipments, leading to logistical disruptions. Similarly, attackers who inject malicious firmware into vehicle control units supplied by an external vendor could enable remote manipulation of vehicle functions. In another scenario, a fleet that unknowingly installs counterfeit microchips in braking systems may experience unexpected failures, creating serious safety hazards.

To mitigate these risks, fleet operators must enforce strict security guidelines for vendors, requiring them to implement encryption, secure software development practices, and penetration testing. Software verification techniques such as cryptographic signing and checksums should be used to detect unauthorized code modifications before deployment. Establishing supply chain auditing procedures ensures that all third-party components meet security standards and are sourced from reputable providers. Working with trusted suppliers who comply with industry security frameworks, such as ISO 21434 for automotive cybersecurity, can further strengthen fleet resilience against cyber threats. By prioritizing these measures, fleet operators can reduce supply chain vulnerabilities and maintain a secure operational environment.

Impact of Compromised Components on Vehicle Security

When supply chain vulnerabilities are exploited, the impact on fleet security can be severe. Modern CMVs depend heavily on electronic systems, meaning that a compromised component can affect multiple aspects of vehicle operation, from data integrity to physical control, creating serious risks for fleet operators.

One of the most critical risks is the potential loss of vehicle control. A compromised ECU, firmware, or software update could allow attackers to override essential driving commands, such as steering, braking, or acceleration, leading to hazardous conditions. Additionally, malicious third-party components can facilitate data breaches, exposing sensitive fleet information, including vehicle locations, cargo details, and operational schedules. For ADS-equipped CMVs, the risks extend further, as these vehicles rely on precise sensor data for decision-making. If LiDAR or radar sensors are tampered with, object detection may be compromised, resulting in unsafe driving behaviors. Furthermore, attackers who exploit supply chain vulnerabilities may be able to remotely disable an entire fleet, causing operational downtime and significant financial losses.

For example, a malicious firmware update sent to a fleet's ADS-equipped trucks could cause them to shut down unexpectedly while on the highway, creating dangerous traffic disruptions. Hackers exploiting vulnerabilities in a third-party vehicle diagnostic system could extract fleet-wide maintenance data and sell it to competitors, giving them an unfair market advantage. Similarly, an attacker who injects a security backdoor into a vehicle's telematics unit could gain remote access to track fleet movements and manipulate routes, leading to potential security breaches or cargo theft.

To mitigate these risks, fleet operators must enforce strict security standards across their supply chain. Vendors should be required to provide security certifications demonstrating compliance with industry best practices. Implementing strict access control policies can prevent unauthorized updates or modifications to vehicle software and hardware, reducing the likelihood of tampering. Conducting independent cybersecurity testing on all third-party hardware and software before deployment ensures that potential vulnerabilities are identified and mitigated early. Additionally, establishing a comprehensive incident response plan enables fleet operators to detect, contain, and respond to supply chain attacks in real time, minimizing disruptions and enhancing overall fleet security. By adopting these measures, fleet operators can strengthen their resilience against supply chain vulnerabilities and safeguard their operations from emerging cybersecurity threats.

HUMAN FACTOR CHALLENGES

While cybersecurity in fleet operations often focuses on technological solutions such as encryption, network security, and intrusion detection, human factors play a significant role in maintaining cybersecurity resilience. Even the most advanced cybersecurity systems can be compromised if personnel are not adequately trained or fail to follow security protocols.

Drivers, fleet operators, and maintenance personnel interact with fleet management systems daily, making them potential targets for cyber threats such as phishing attacks, social engineering, and credential theft. A lack of cybersecurity awareness and improper security practices can lead to severe vulnerabilities, potentially allowing attackers to gain unauthorized access to fleet operations.

COGNITIVE WORKLOAD CONSIDERATIONS

Managing cybersecurity in mixed fleets presents cognitive challenges for operators who oversee both traditional and automated vehicles. The complexity of monitoring cybersecurity threats, responding to alerts, and maintaining situational awareness can significantly increase cognitive workload, affecting decision-making and overall system security. As fleet operations incorporate advanced digital interfaces, operators must balance attention between multiple systems, increasing the risk of overload, stress, and fatigue.

Cognitive workload is particularly relevant in cybersecurity tasks that require continuous vigilance. Operators managing cybersecurity for fleet vehicles must detect anomalies, verify system integrity, and respond to security incidents while coordinating with other operational responsibilities. Increased cognitive demands can lead to delayed threat detection, misinterpretation of security alerts, and higher susceptibility to social engineering attacks. Stress and fatigue further compound these risks, reducing an operator's ability to process information efficiently and make sound security decisions.

Fatigue-related cognitive impairment is especially concerning for cybersecurity personnel responsible for teleoperation, real-time network monitoring, and fleet management. Long shifts, high-alert work environments, and the need for rapid decision-making can diminish vigilance, increasing the likelihood of overlooked security breaches or delayed responses to cyber incidents. Designing cybersecurity systems with human cognitive limitations in mind is essential to maintaining fleet security and reducing the risks associated with mental overload.

To address these challenges, cybersecurity interfaces should be designed using human-centered principles that support efficient information processing and decision-making. Security alerts should be prioritized based on risk severity, minimizing unnecessary notifications that contribute to cognitive overload. Adaptive automation can also help reduce workload by automating routine security checks while ensuring that operators retain control over critical decision-making processes.

Implementing workload management strategies, such as shift rotations, scheduled breaks, and workload-sharing mechanisms, can help alleviate stress and improve operator performance. Cybersecurity training programs should also incorporate cognitive workload considerations, preparing operators to manage security tasks effectively under varying conditions. By designing security interfaces that align with human cognitive capabilities and implementing workload management strategies, fleet operators can enhance cybersecurity resilience while reducing the risks associated with operator fatigue and overload.

Training Gaps in Cybersecurity Awareness

Many cybersecurity incidents in fleet operations result from human errors rather than direct system vulnerabilities. A significant challenge in mixed fleet security is the lack of comprehensive cybersecurity training for personnel who operate and manage CMVs, including automated and connected trucks.

Common training gaps:

- Limited understanding of cybersecurity threats – Many fleet employees are not familiar with cyber risks such as phishing, malware, or unauthorized access attempts, making them more susceptible to attacks.
- Inconsistent security policies across fleet operations – Fleet personnel may not follow standardized security procedures due to a lack of clear cybersecurity policies or inconsistent enforcement.
- Failure to recognize suspicious activities – Employees who are unaware of cybersecurity best practices may overlook warning signs of a cyberattack, such as unusual system behavior, unauthorized access requests, or fake software updates.
- Weak credential management – Poor password hygiene, such as reusing credentials or failing to update passwords regularly, can increase the risk of credential-based attacks.
- Lack of secure device usage practices – Drivers and operators often use personal devices to access fleet management systems. Without proper cybersecurity training, they may connect to unsecured Wi-Fi networks or download unverified applications that introduce security vulnerabilities.

Example threats:

- A driver unknowingly clicks on a phishing email disguised as a routine fleet management update, compromising their login credentials.
- A fleet technician installs unauthorized third-party software on a CMV diagnostic system, introducing a security backdoor that attackers exploit.
- A dispatcher shares login credentials with multiple employees, increasing the risk of unauthorized system access.

Mitigation strategies:

- Implement regular cybersecurity training programs for fleet personnel, focusing on recognizing cyber threats and secure system interactions.
- Establish clear security policies for credential management, including password complexity requirements and MFA.
- Conduct simulated cyberattacks (e.g., phishing tests) to assess employee awareness and reinforce best practices.
- Require all employees to complete cybersecurity certification courses tailored for fleet operations.
- Develop an incident response plan that educates personnel on how to report and respond to potential cybersecurity breaches.

Risks Associated with Driver/Operator Behavior in Fleet Security

Human errors and risky behaviors among fleet personnel can introduce vulnerabilities in cybersecurity, even when strong technical security measures are in place. Drivers and operators may unintentionally create security gaps by mishandling sensitive data, using unsecured communication channels, or neglecting cybersecurity protocols. Figure 9.2 is a threat radar for fleet cybersecurity based on different job roles, it illustrates how each job role poses different types of cybersecurity threats on different risk levels for fleets. For example, drivers can potentially cause medium risk in Vehicle Control & Safety, Insider Threats & Human Factors, and Infrastructure & Operational Disruption, and cause low risk in Data Breach & Privacy and Supply Chain & Third-Party Risks. Different fleets can define each job role differently, and this graph has been made based on the generalized nature of each job role in the industry. This graph is only made to assist with the

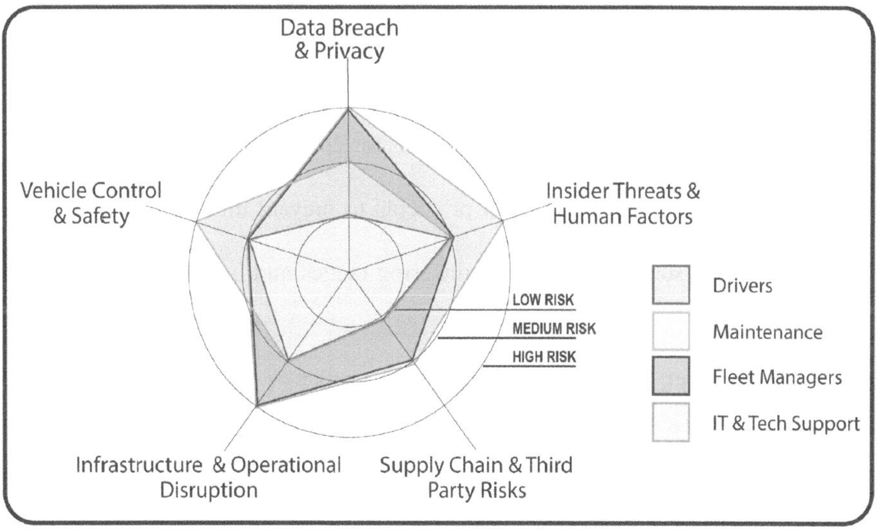

FIGURE 9.2 Fleet cybersecurity threat radar.

understanding of the relationship between threat levels and job roles in fleets; this is not from a result of scientific study.

Common security risks caused by operator behavior:

- Use of unauthorized devices – Drivers may plug personal USB drives, smartphones, or other unapproved devices into fleet management systems, potentially introducing malware.
- Failure to log out of systems – Operators who leave fleet terminals or onboard computer systems logged in risk unauthorized access by malicious actors.
- Connecting to unsecured networks – Drivers often use public Wi-Fi at truck stops, rest areas, or hotels, increasing the risk of network interception and data theft.
- Bypassing security features for convenience – Operators may disable firewalls, antivirus software, or GPS tracking to improve device performance, unknowingly exposing fleet systems to cyber threats.
- Social engineering attacks – Attackers may impersonate fleet managers, IT personnel, or regulatory officials to trick drivers into providing sensitive fleet credentials or downloading malicious software.

Example threats:

- A driver unknowingly connects their fleet tablet to an unsecured Wi-Fi network at a truck stop, allowing a hacker to intercept fleet communications.
- An operator clicks on a fraudulent pop-up warning about a system update, unknowingly downloading malware that compromises fleet data.
- A fleet dispatcher shares operational details over an unencrypted messaging app, exposing shipment schedules to potential cybercriminals.

Mitigation strategies:

- Prohibit the use of personal devices on fleet management networks and enforce strict access control policies.
- Require fleet personnel to log out of all terminals and onboard systems after each use.
- Provide secure mobile network solutions for fleet communication, reducing reliance on public Wi-Fi.
- Establish strong authentication protocols to prevent unauthorized system modifications.
- Educate personnel on social engineering tactics and how to verify legitimate cybersecurity requests.

BEST PRACTICES FOR ENHANCING CYBERSECURITY IN GROUND TRANSPORTATION

The increasing reliance on automated and connected technologies in CMV fleets has made cybersecurity a top priority for fleet operators. As cyber threats evolve, fleets must adopt comprehensive cybersecurity strategies to safeguard data, vehicle systems, and overall operational integrity.

Cybersecurity best practices in transportation focus on protecting network infrastructure, securing data, training personnel, enforcing physical security, implementing incident response plans, and continuously monitoring for threats. By integrating these strategies, fleet operators can mitigate cyber risks and ensure the safe operation of mixed fleets, including both conventional and ADS-equipped vehicles.

RBAC for Fleet Management

RBAC ensures that employees have access only to the data and system functions necessary for their job responsibilities. Implementing RBAC reduces the risk of accidental system modifications, unauthorized data exposure, and privilege escalation attacks.

Fleet management systems handle multiple operational tasks, including vehicle dispatching, diagnostics, compliance monitoring, and maintenance scheduling. Without a structured access control policy, employees may inadvertently access critical system functions beyond their scope of work, leading to potential security risks.

Best practices for implementing RBAC:

- Define specific user roles such as drivers, dispatchers, maintenance personnel, and system administrators, each with predefined access privileges.
- Restrict administrative privileges to a limited number of authorized personnel, preventing unauthorized modifications to system configurations and cybersecurity settings.
- Implement least privilege principles, ensuring that users can only perform actions required for their job roles while preventing access to unnecessary system features.
- Regularly review and update role assignments to reflect changes in personnel responsibilities, preventing former employees from retaining access to fleet systems.
- Log and monitor user activities to detect unauthorized access attempts or privilege misuse within the fleet management platform.

CYBERSECURITY TRAINING FOR FLEET PERSONNEL

Fleet cybersecurity is not only a technological challenge but also a human factor issue. Drivers, operators, dispatchers, and maintenance personnel play a significant role in maintaining cybersecurity resilience within fleet operations. Even with advanced security measures in place, human errors, lack of awareness, and inadequate training can expose fleet management systems to cyber threats. Ensuring that all personnel are educated on cybersecurity risks and best practices is essential to preventing unauthorized access, data breaches, and system manipulation.

Cybersecurity training programs must be designed to address real-world risks encountered by fleet personnel. These programs should emphasize cybersecurity awareness, secure handling of digital systems, and compliance with cybersecurity policies to reduce the risk of human-related security incidents.

Importance of Cybersecurity Awareness among Drivers and Operators

Drivers and fleet operators interact with multiple digital systems daily, including telematics devices, mobile applications, electronic logging devices (ELDs), and fleet management platforms. If they are not aware of cybersecurity threats, they may unknowingly expose fleet systems to risks such as phishing attacks, malware infections, or unauthorized access attempts.

Common cybersecurity risks for fleet personnel:

- Falling victim to phishing attacks that attempt to steal login credentials or distribute malware.
- Using weak or reused passwords that can be easily guessed or exploited.
- Connecting fleet devices to unsecured public Wi-Fi networks, increasing the risk of network interception.
- Failing to log out of shared fleet management systems, allowing unauthorized access.
- Plugging unauthorized USB devices into fleet telematics systems, potentially introducing malware.

Best practices for cybersecurity awareness training:

- Educate personnel on recognizing phishing attempts, suspicious emails, and fraudulent system alerts.
- Provide clear guidelines on password management, emphasizing the use of strong, unique passwords and MFA.
- Instruct fleet personnel to avoid using public Wi-Fi networks for logging into fleet systems unless connected through a secure VPN.
- Encourage drivers and operators to report any unusual system activity, suspected cyber threats, or unauthorized login attempts.
- Implement regular refresher courses to ensure that personnel stay informed about evolving cybersecurity risks.

Cybersecurity awareness is a critical component of fleet security. Without proper training, personnel may inadvertently expose fleet systems to cyber threats, making continuous education and reinforcement of cybersecurity best practices essential for maintaining secure operations.

Regular Training Programs and Certification for Cybersecurity Protocols

To ensure a consistent level of cybersecurity readiness across fleet operations, structured training programs and certification requirements should be implemented. Cybersecurity training must be integrated into onboarding procedures for new employees and maintained as an ongoing requirement for all personnel who interact with fleet management systems.

Elements of an effective cybersecurity training program:

- Cybersecurity fundamentals, including recognizing threats, secure login practices, and safe internet usage.
- Hands-on training with real-world cybersecurity scenarios, such as phishing simulations and incident response exercises.
- Clear explanations of fleet security policies, compliance requirements, and the role of each employee in maintaining cybersecurity.
- Step-by-step procedures for reporting security incidents and responding to potential cyber threats.

Best practices for cybersecurity training and certification:

- Require all fleet personnel to complete a cybersecurity certification program as part of their onboarding process.
- Conduct annual cybersecurity refresher courses to keep employees informed about new threats and security updates.
- Implement role-specific cybersecurity training, ensuring that dispatchers, drivers, and IT personnel receive tailored education relevant to their responsibilities.
- Perform cybersecurity awareness assessments to evaluate personnel knowledge and reinforce key concepts.
- Use interactive e-learning platforms to provide accessible cybersecurity training materials, allowing personnel to complete training at their own pace.

By integrating regular cybersecurity training and certification into fleet operations, organizations can create a culture of cybersecurity awareness and reduce the risk of security breaches caused by human error. Ensuring that personnel understand their role in cybersecurity enhances the overall protection of fleet systems and strengthens the resilience of transportation networks against cyber threats.

Learning Principles and Culture Building

Effective cybersecurity training for fleet personnel goes beyond routine compliance requirements. To ensure lasting security awareness and behavioral change, training programs must incorporate adult learning principles, scenario-based learning techniques, and strategies for fostering a security-conscious culture. By tailoring cybersecurity education to the needs of adult learners, fleet operators can improve retention, engagement, and the practical application of security knowledge.

Adult Learning Principles in Cybersecurity Training

Adult learners benefit from training that is relevant, self-directed, and experience based. Traditional lecture-style cybersecurity training may not be effective in engaging fleet personnel, as passive learning methods often fail to translate into real-world security behaviors. Instead, cybersecurity training should incorporate:

- Relevance and context – Training materials should be directly applicable to the daily responsibilities of fleet personnel, demonstrating how cybersecurity threats impact vehicle operations, telematics systems, and data security.
- Active participation – Learners should be encouraged to engage in hands-on activities, simulations, and real-world cybersecurity exercises rather than passively absorbing information.
- Self-directed learning opportunities – Employees should have access to on-demand cybersecurity training resources, allowing them to review critical concepts at their own pace.
- Feedback and reinforcement – Regular assessments, knowledge checks, and refresher courses should be used to reinforce cybersecurity principles and keep personnel up to date on emerging threats.

Simulation and Scenario-Based Learning Approaches

Hands-on learning through simulations and scenario-based training enhances cybersecurity preparedness by exposing personnel to realistic security threats in a controlled environment. Rather than relying solely on theoretical instruction, interactive learning techniques provide personnel with direct experience in identifying and responding to cyber incidents.

Effective simulation-based cybersecurity training includes:

- Phishing attack simulations – Employees receive simulated phishing emails to assess their ability to recognize and report social engineering threats.
- Incident response drills – Fleet operators conduct cybersecurity exercises where personnel practice responding to simulated ransomware attacks, unauthorized access attempts, or data breaches.
- Cybersecurity escape rooms or challenge-based learning – Teams work together to solve cybersecurity challenges, reinforcing key security principles through problem-solving and collaboration.
- Hands-on vulnerability testing – IT and cybersecurity personnel engage in penetration testing exercises to identify weaknesses in fleet management systems.

By incorporating scenario-based learning, fleet personnel gain practical cybersecurity skills that prepare them for real-world cyber threats. These exercises help build confidence in recognizing and responding to cyber incidents, reducing the likelihood of human errors that could compromise fleet security.

Building a Security Culture beyond Compliance Training

A strong security culture is essential for sustaining cybersecurity awareness beyond formal training sessions. Organizations must foster an environment where cybersecurity is viewed as an integral part of fleet operations rather than a one-time compliance requirement. This requires continuous engagement, leadership support, and integration of security best practices into daily workflows.

Key strategies for building a security culture include:

- Leadership involvement – Fleet managers and executives should actively promote cybersecurity best practices and lead by example in following security protocols.
- Regular security briefings – Short, frequent discussions on emerging cybersecurity threats, recent security incidents, and best practices help keep cybersecurity top of mind.
- Peer-led security training – Experienced personnel can mentor new employees, reinforcing cybersecurity principles through team-based learning.
- Gamification and incentives – Organizations can introduce cybersecurity challenges, reward employees for successfully completing security training modules, and recognize personnel who demonstrate strong security practices.
- Integration with operational routines – Cybersecurity protocols should be seamlessly incorporated into daily fleet operations, ensuring that employees follow security procedures as part of their regular duties rather than as an isolated requirement.

By applying adult learning principles, leveraging scenario-based learning, and fostering a strong security culture, fleet operators can ensure that cybersecurity training is both effective and sustainable. Building a workforce that is continuously engaged in cybersecurity awareness strengthens fleet security and reduces the likelihood of human-related security incidents.

PHYSICAL SECURITY MEASURES

Cybersecurity in fleet operations extends beyond digital threats to include physical security risks. Unauthorized physical access to vehicles, diagnostic tools, and control systems can compromise cybersecurity measures, allowing attackers to manipulate vehicle functions, install malicious hardware, or extract sensitive fleet data. Securing the physical components of fleet management systems and ADS is essential to maintaining cybersecurity integrity.

Physical security measures help prevent unauthorized individuals from accessing vehicle hardware, software interfaces, and networked components. Implementing strict physical security protocols reduces the risk of cyber-physical attacks that could disrupt fleet operations or compromise data confidentiality.

Restricting Unauthorized Access to Vehicle Hardware and Software

Fleet vehicles, especially ADS-equipped CMVs, contain critical electronic components that must be protected against unauthorized access. Attackers who gain physical access to these systems may attempt to modify firmware, install tracking devices, or manipulate control units to introduce security vulnerabilities.

Common risks associated with unauthorized access:

- Direct tampering with ECUs, OBD ports, or telematics systems.
- Unauthorized personnel gaining access to fleet terminals, loading docks, or maintenance facilities.

- Physical installation of rogue devices, such as GPS jammers, malware-infected USB drives, or wireless sniffing equipment.
- Theft or unauthorized removal of fleet tablets, laptops, or mobile devices containing sensitive fleet data.

Best practices for restricting unauthorized access:

- Secure vehicle access points by using electronic locks, biometric authentication, or keyless entry systems that require authorization.
- Implement tamper-evident security seals on diagnostic ports, ECUs, and telematics units to detect unauthorized access attempts.
- Enforce strict access control measures in fleet yards, garages, and maintenance centers by requiring identification and entry authorization.
- Prohibit the use of unauthorized USB devices and external storage media on fleet systems to prevent malware installation.
- Regularly inspect fleet vehicles for signs of tampering, including unexpected modifications to communication modules or diagnostic tools.

Unauthorized access to vehicle hardware and software can create significant security vulnerabilities. By implementing strict access control measures and securing physical entry points, fleet operators can reduce the risk of cyber-physical attacks and unauthorized system modifications.

Tamper-Proof Designs for ADS and Fleet Control Systems

ADS-equipped vehicles and connected fleet systems require robust tamper-proof designs to protect critical hardware components from external manipulation. Attackers who gain access to ADS components may attempt to interfere with sensor data, disable automated functions, or inject false information into vehicle control systems.

Key vulnerabilities in ADS and fleet control systems:

- Physical obstruction, damage, or repositioning of sensors such as LiDAR, radar, and cameras, leading to incorrect vehicle perception and unsafe driving behavior.
- Unauthorized access to ADS computing units, which could allow attackers to modify algorithms, disable safety features, or introduce malware into autonomous driving software.
- Manipulation of GPS antennas and vehicle-to-everything (V2X) communication modules to disrupt navigation and fleet coordination.
- Interference with fleet control centers, where unauthorized personnel could modify dispatch instructions or reroute shipments using compromised access points.

Best practices for tamper-proof designs:

- Secure all ADS sensors and communication modules with protective enclosures to prevent physical damage or repositioning.
- Implement hardware encryption for critical ADS computing units to prevent unauthorized firmware modifications.

- Deploy redundant sensor arrays to cross-verify perception data, reducing the impact of intentional sensor manipulation.
- Use anti-tamper coatings and shielding for vehicle wiring and networked components to protect against unauthorized physical modifications.
- Design fleet control centers with restricted access policies, requiring biometric authentication or secure entry badges for personnel managing fleet operations.

By implementing tamper-proof designs, fleet operators can enhance the resilience of ADS-equipped vehicles and protect against physical security threats. Preventing unauthorized access to critical hardware ensures that automated functions operate as intended and that fleet security remains uncompromised.

FAILURE AND RECOVERY STRATEGIES

Cybersecurity threats in fleet operations are not always preventable, making it essential to have failure and recovery strategies in place to minimize disruption and ensure the safety of both vehicles and personnel. A well-defined response plan enables fleet operators to contain cyber incidents, restore systems efficiently, and mitigate the impact of attacks on vehicle operations.

In mixed fleets, failure and recovery strategies must account for both traditional and ADS-equipped CMVs. Ensuring that ADS-equipped vehicles can enter a minimal risk condition (MRC) in the event of a cyber breach is critical to maintaining road safety.

Developing Incident Response Plans for Cyber Breaches

A cybersecurity incident response plan provides a structured approach to detecting, containing, and mitigating cyberattacks. Without a well-prepared plan, fleet operators risk prolonged system downtime, data breaches, and safety hazards.

Key components of an effective incident response plan:

- Threat detection and identification: Establishing real-time monitoring tools to identify unauthorized access attempts, data breaches, or system anomalies.
- Containment procedures: Defining immediate actions to limit the spread of cyber threats, such as isolating compromised systems or disabling affected network connections.
- Response team coordination: Assigning responsibilities to key personnel, including IT security teams, fleet managers, and legal advisors, to ensure a swift response.
- Communication protocols: Developing guidelines for notifying internal stakeholders, regulatory agencies, and law enforcement if a cyber breach affects fleet security.
- Data recovery and system restoration: Implementing backup strategies to restore lost or corrupted data, ensuring fleet operations resume with minimal disruption.

- Post-incident analysis: Reviewing the root causes of cyber incidents, identifying vulnerabilities, and updating security policies to prevent future breaches.

Best practices for incident response implementation:

- Conduct regular cybersecurity drills to test the effectiveness of response plans and improve team preparedness.
- Establish a cyber incident reporting system for fleet personnel to report suspicious activities.
- Maintain secure, offline backups of fleet management data to facilitate rapid recovery after a cyberattack.
- Document all cyber incidents and lessons learned to refine future response strategies.

An effective incident response plan ensures that fleet operators can quickly contain cybersecurity threats and restore normal operations with minimal disruption. Without a proactive strategy, cyberattacks can lead to significant financial and operational losses.

Ensuring ADS-Equipped Vehicles Enter a MRC

ADS-equipped vehicles rely on automated driving functions that must be safeguarded against cyber intrusions. If a cybersecurity incident affects an ADS-equipped CMV, the system must be able to transition into a MRC to prevent accidents and maintain road safety.

MRC refers to a predefined safe state that an ADS-equipped vehicle enters when an unexpected system failure occurs. In the event of a cyber breach, vehicles must be programmed to take appropriate safety measures to minimize risks for passengers, other road users, and fleet operators.

Key considerations for MRC implementation in ADS-equipped fleets:

- Safe stopping mechanisms: Vehicles should be able to gradually slow down and come to a controlled stop in a safe location, such as the shoulder of a highway or a designated emergency zone.
- Automated alerts: The system should notify fleet operators, emergency services, and nearby vehicles about the cyber-related failure, providing real-time status updates.
- Fallback operational modes: If a cyber intrusion disables core ADS functionalities, the system should default to a limited driving mode that allows the vehicle to continue moving safely under human supervision.
- Redundant safety systems: Backup control mechanisms, including redundant communication channels and fail-safe braking systems, should be in place to mitigate the effects of cyberattacks.

Best practices for MRC implementation:

- Conduct regular cybersecurity stress tests on ADS-equipped vehicles to evaluate system responses under simulated attack conditions.

- Ensure that MRC protocols comply with industry safety regulations and align with government cybersecurity standards for automated vehicles.
- Integrate secure remote shutdown capabilities that allow fleet operators to safely disable compromised ADS functions if necessary.
- Develop a decision-making hierarchy for ADS failure responses, prioritizing safety, system integrity, and fleet coordination.

Ensuring that ADS-equipped vehicles can enter a MRC during a cyberattack reduces the likelihood of system failures leading to accidents or fleet disruptions. As automated driving technology advances, robust MRC strategies will be essential for securing mixed fleets against evolving cyber threats.

CONTINUOUS MONITORING AND PENETRATION TESTING

Cyber threats to fleet operations are constantly evolving, requiring ongoing vigilance to detect and respond to potential attacks before they can cause harm. Continuous monitoring and penetration testing are essential cybersecurity practices that help fleet operators identify vulnerabilities, track suspicious activities, and assess the effectiveness of existing security controls.

Real-time monitoring provides immediate insights into security threats, allowing fleet operators to take proactive measures against unauthorized access or system anomalies. Regular cybersecurity audits and penetration testing ensure that fleet systems remain resilient against emerging threats and evolving attack techniques.

Real-Time Threat Monitoring in Fleet Operations

Real-time threat monitoring is critical for identifying and mitigating cyber risks before they escalate. Fleet management systems, telematics networks, and ADS-equipped vehicles generate vast amounts of operational data, making it essential to have automated monitoring tools that can detect irregularities.

Continuous monitoring involves tracking network activity, system access logs, and vehicle diagnostics for signs of unauthorized activity. Any deviations from normal operational patterns can signal a potential cybersecurity incident, such as an intrusion attempt, malware infection, or unauthorized remote access.

Key considerations for real-time threat monitoring include:

- Using IDS and intrusion prevention systems to identify cyber threats in real time.
- Implementing automated alert mechanisms that notify fleet security teams of unusual login attempts, unexpected data transfers, or unauthorized software changes.
- Establishing centralized security information and event management systems to collect and analyze security event logs from multiple fleet systems.

- Monitoring network traffic for anomalies, such as data spikes, unrecognized IP addresses, or irregular communication patterns between vehicles and fleet management servers.
- Deploying endpoint detection and response solutions on fleet management devices to prevent malware execution and detect advanced cyber threats.

Proactive threat monitoring allows fleet operators to respond quickly to security incidents, reducing the risk of data breaches, fleet disruptions, or vehicle hijacking attempts. Implementing real-time monitoring solutions helps secure fleet networks and ensure the integrity of digital fleet operations.

Importance of Regular Cybersecurity Audits

Regular cybersecurity audits and penetration testing help fleet operators identify and address security weaknesses before cybercriminals can exploit them. Unlike real-time monitoring, which focuses on detecting active threats, cybersecurity audits provide a structured review of existing security policies, network configurations, and software vulnerabilities.

Penetration testing involves simulating cyberattacks against fleet systems to assess how well security defenses hold up against real-world threats. These tests can reveal gaps in access control mechanisms, software patching practices, and employee cybersecurity awareness.

Key elements of effective cybersecurity audits include:

- Reviewing access control policies to ensure only authorized personnel have system privileges.
- Evaluating encryption protocols used for data storage and transmission to verify compliance with security standards.
- Identifying outdated software, unpatched vulnerabilities, or misconfigured network settings that could be exploited by attackers.
- Assessing cybersecurity training programs to ensure fleet personnel are prepared to recognize and respond to cyber threats.
- Examining third-party vendor security policies to prevent supply chain vulnerabilities.

Best practices for penetration testing include:

- Conducting red team[5] exercises, where ethical hackers attempt to breach fleet security defenses to uncover weaknesses.
- Testing the resilience of ADS-equipped CMVs against remote hacking attempts or GPS spoofing attacks.
- Simulating phishing campaigns to measure how well employees recognize and avoid social engineering tactics.
- Performing wireless security assessments to ensure fleet telematics systems are protected against eavesdropping and unauthorized access.

Regular cybersecurity audits and penetration testing provide valuable insights into fleet security posture, helping fleet operators strengthen defenses against cyber threats. By continuously evaluating and improving cybersecurity measures, fleet operations can remain resilient in the face of evolving cyber risks.

CYBERSECURITY CONSIDERATIONS FOR SPECIALIZED ADS USE CASES

As ADS continue to advance, cybersecurity risks extend beyond individual vehicles to encompass networked transportation ecosystems. Specialized ADS use cases such as cooperative perception, teleoperation, vehicle platooning, and regulatory interactions introduce unique cybersecurity challenges. These scenarios rely on complex data-sharing architectures, remote access capabilities, and multi-vehicle coordination, all of which expand the attack surface for cyber threats.

Ensuring the cybersecurity of these specialized use cases requires robust encryption, secure authentication, and proactive threat monitoring. The following sections explore key cybersecurity considerations in cooperative perception, teleoperation, automated truck convoys, and interactions with regulatory authorities.

COOPERATIVE PERCEPTION AND V2X COMMUNICATION

Cooperative perception enables ADS-equipped vehicles to share real-time sensor data with other road users and infrastructure, improving situational awareness and traffic efficiency. V2X communication facilitates this data exchange by allowing vehicles to communicate with each other, roadside units, traffic control centers, and even pedestrians. However, the introduction of these communication channels also presents significant cybersecurity risks that must be addressed to prevent malicious interference.

One of the primary concerns in cooperative perception is the integrity of data shared between vehicles and infrastructure. Cybercriminals can manipulate or intercept these data streams, introducing false information that misleads ADS-equipped vehicles and disrupts traffic operations. Transmission of falsified sensor data can cause vehicles to misinterpret road conditions, leading to unsafe driving decisions. Spoofing attacks present another major risk, as attackers can create phantom vehicles or fake road hazards within ADS sensor networks, prompting unnecessary evasive maneuvers or traffic congestion. Additionally, unauthorized access to V2X communication channels can allow malicious actors to inject misleading messages, further compromising vehicle decision-making. DoS attacks pose an additional threat by overwhelming V2X networks, disrupting the timely exchange of critical data and degrading system reliability.

To secure cooperative perception and V2X communication, robust cybersecurity measures must be implemented. Cryptographic authentication should be applied to all transmitted data to ensure message integrity and prevent unauthorized modifications. Digital signatures can further enhance security by verifying the legitimacy of V2X messages and preventing attackers from injecting false information into the network. Deploying anomaly detection algorithms is essential for identifying and rejecting fraudulent sensor data before it affects vehicle operations. Additionally, secure communication protocols such as IEEE 1609.2 should be used to encrypt

V2X transmissions, ensuring that only authorized entities can access and exchange critical sensor data. By adopting these best practices, fleet operators and transportation authorities can enhance the security of cooperative perception, ensuring safe and reliable data-sharing between ADS-equipped vehicles and infrastructure.

Teleoperation and Remote Assistance Risks

Teleoperation and remote assistance allow human operators to intervene in ADS-equipped vehicle operations when automated systems reach their operational limits, such as navigating construction zones or responding to emergency situations. While this capability enhances operational flexibility, it also introduces significant cybersecurity risks related to remote access, data transmission security, and the potential for unauthorized control over vehicles.

Remote vehicle control relies on high-bandwidth, low-latency communication channels to ensure safe and precise operations. If an attacker gains access to these systems, they could intercept control signals, disrupt fleet monitoring, or even take unauthorized control of a vehicle, posing severe safety and operational risks. One of the primary cybersecurity threats is man-in-the-middle attacks, where an attacker intercepts and alters remote driving commands, potentially leading to dangerous vehicle maneuvers. Weak authentication mechanisms further increase the risk of unauthorized access to teleoperation platforms, allowing malicious actors to manipulate fleet operations. Additionally, jamming or DoS attacks could disrupt teleoperation connectivity, causing vehicles to lose remote guidance and enter unsafe conditions. Another critical vulnerability is the exploitation of unpatched software, where attackers inject malicious commands into the teleoperation system, compromising vehicle safety.

To mitigate these risks, robust cybersecurity measures must be implemented. Enforcing MFA for all remote operators accessing teleoperation platforms ensures that only authorized personnel can issue vehicle commands. Encrypting all control signals using strong encryption protocols prevents unauthorized interception and manipulation of vehicle operations. Continuous monitoring and real-time alerting should be deployed to detect and respond to anomalous remote control activity, allowing fleet operators to quickly identify potential cyber intrusions. Additionally, establishing automatic fail-safe mechanisms is essential – ADS-equipped vehicles should be programmed to transition into a MRC if remote connectivity is lost, preventing unsafe situations from escalating. By integrating these security best practices, fleet operators can enhance the resilience of teleoperation systems while maintaining safe and reliable vehicle operations.

Platooning and Automated Truck Convoys

Platooning technology enables multiple trucks to travel in close formation with synchronized acceleration, braking, and steering. By leveraging V2V communication, platooning enhances fuel efficiency and reduces traffic congestion. However, this high level of automation and inter-vehicle coordination introduces cybersecurity challenges that could impact the safety and stability of the convoy.

One of the most critical risks in platooning is the potential for unauthorized access and control over multiple vehicles. If an attacker gains entry into the platoon's control network, they could manipulate the convoy's behavior, leading to traffic disruptions or even collisions. Cyber threats to platooning operations include unauthorized takeover of the lead vehicle, which could impact all following trucks, and spoofing or jamming of V2V communication, causing vehicles to respond to false data inputs. Attackers may also inject malicious commands remotely, forcing vehicles to decouple from the platoon or engage in sudden emergency braking. Additionally, vulnerabilities in platooning algorithms could be exploited, leading to unpredictable vehicle behavior and compromising the overall stability of the convoy.

To mitigate these cybersecurity risks, robust security measures must be implemented. Encrypted V2V communication is essential to protect data exchanges within the platoon, ensuring that only authenticated signals are processed. Strong authentication and access control mechanisms should be enforced to prevent unauthorized vehicles from joining the convoy. Regular security audits and penetration testing should be conducted on platooning software to identify and address potential vulnerabilities before they can be exploited. Furthermore, deploying redundant perception systems allows each vehicle to independently verify platooning commands, reducing reliance on a single point of failure. By integrating these security best practices, fleet operators can enhance the resilience of platooning technology while maintaining safe and efficient convoy operations.

CYBERSECURITY MEASURES FOR ROADSIDE INSPECTIONS AND LAW ENFORCEMENT INTERACTIONS

ADS-equipped vehicles and connected fleets must comply with roadside inspections and regulatory enforcement procedures, requiring data exchanges between fleet operators, law enforcement agencies, and regulatory authorities. Without proper security measures, these interactions can expose sensitive fleet information to cyber threats, making it essential to implement robust data protection strategies.

One of the primary concerns during inspections is protecting sensitive data from unauthorized access. Fleet operators are required to share operational data, driver logs, and compliance records with roadside inspectors, but if these exchanges are not properly secured, confidential fleet information could be exposed. Cybersecurity risks include unauthorized access to fleet data through compromised inspection terminals, interception of wireless data transmissions during regulatory checks, and data tampering that falsifies compliance records or obscures fleet violations. To mitigate these threats, all data exchanges between fleet systems and roadside inspection terminals should be encrypted to prevent interception. Strict access control policies must be enforced to ensure that only authorized personnel can retrieve compliance records. Additionally, blockchain-based audit trails can be implemented to maintain the integrity of inspection logs and prevent unauthorized modifications.

Another significant security risk arises from unauthorized network access during regulatory checks. Roadside inspections often require connectivity to fleet telematics systems, ELDs, and vehicle diagnostic interfaces. If these connections are not properly secured, cybercriminals could exploit them to infiltrate fleet networks.

Wireless sniffing attacks, for example, could capture unencrypted data transmitted during inspections, while compromised regulatory access points could be used to inject malicious software into fleet systems. Additionally, attackers could manipulate vehicle configurations by exploiting vulnerabilities in diagnostic tools used during inspections.

To prevent unauthorized network access, all roadside inspection data transmissions should be conducted over encrypted channels, ensuring that sensitive fleet information remains protected. Fleet operators must also verify the authenticity of law enforcement and regulatory personnel before granting access to fleet systems to prevent unauthorized intrusions. Isolating fleet telematics systems from external networks is another crucial measure, as it minimizes the risk of unauthorized access during inspections. By implementing these best practices, fleet operators can enhance the security of regulatory interactions and protect their vehicles from cyber threats while ensuring compliance with inspection procedures.

HUMAN CENTERED CYBERSECURITY IN FLEET OPERATIONS

Cybersecurity in fleet management must be designed with the end user in mind. While strong security measures are essential to protecting vehicles, data, and operations, overly complex or intrusive security protocols can create frustration, inefficiencies, and even unintentional security breaches. Human-centered cybersecurity, also known as user-centered security design, ensures that security measures support operators' tasks rather than obstructing them. This approach balances security requirements with usability, ensuring that cybersecurity practices align with how drivers, fleet managers, and maintenance personnel interact with digital systems.

Designing Security Measures That Do Not Impede Operators' Primary Tasks

Fleet operators and drivers must manage a variety of operational tasks, including vehicle monitoring, route adjustments, cargo tracking, and safety compliance. If security measures are too cumbersome, they can slow down operations, increase cognitive workload, or encourage users to bypass them altogether.

Best practices for designing non-intrusive security measures include:

- Minimizing authentication friction – Implementing secure yet convenient authentication methods, such as biometric login or single sign-on, reduces login fatigue while maintaining security.
- Automating routine security checks – Background security processes, such as automatic software updates and real-time threat monitoring, should run without requiring constant user intervention.
- Streamlining access control mechanisms – RBAC should ensure that employees have access only to the tools they need, without adding unnecessary layers of authentication for routine tasks.

- Ensuring security compliance integrates naturally into workflow – Security policies should align with existing operational procedures, so that following cybersecurity protocols does not feel like an extra burden.

Balancing Security Requirements with Usability Needs

A common challenge in cybersecurity is balancing security with usability. Security measures that are too strict or complex can lead to user frustration and workarounds that introduce new vulnerabilities. For example, if password requirements are too rigid, users may write them down or reuse them across multiple systems, increasing security risks.

To achieve a balance between security and usability, security designs should:

- Provide clear and intuitive interfaces – Security settings and procedures should be easy to understand and navigate, reducing the need for extensive training or support.
- Avoid unnecessary authentication barriers – Using adaptive authentication, which adjusts security measures based on risk levels, can prevent unnecessary security prompts while still protecting sensitive data.
- Support error recovery – If a user makes a mistake, such as entering the wrong login credentials, the system should provide clear guidance on how to resolve the issue rather than locking them out immediately.
- Allow user feedback – Operators should have a way to report usability concerns related to security features, helping designers refine security implementations to better fit real-world operational needs.

Creating Alerts and Notifications That Capture Attention without Causing Alarm Fatigue

Effective security alerts are critical for cybersecurity awareness, but excessive or poorly designed notifications can lead to alarm fatigue, where users begin ignoring or dismissing important warnings. If operators are overwhelmed with frequent or unclear alerts, they may fail to respond appropriately to real security threats.

Key principles for designing effective cybersecurity alerts include:

- Prioritization and filtering – Security notifications should be categorized by severity, ensuring that critical threats take precedence over low-priority messages.
- Clear and actionable messages – Alerts should provide specific instructions on what action to take, rather than vague warnings that create uncertainty.
- Minimizing unnecessary alerts – Reducing redundant or low-relevance notifications helps keep operators focused on real threats.
- Using multiple alert modalities – Combining visual, auditory, and haptic alerts can improve awareness while reducing reliance on a single notification method.

- Context-sensitive alerts – Security warnings should be relevant to the operator's current task, reducing disruptions while ensuring timely response to threats.

By incorporating human-centered cybersecurity principles, fleet operators can create security measures that enhance protection without interfering with daily operations. A well-designed cybersecurity system supports users by integrating security seamlessly into workflows, improving compliance, and ensuring that critical alerts capture attention without overwhelming personnel. This approach strengthens cybersecurity while maintaining operational efficiency in modern fleet environments.

FUTURE TRENDS AND EVOLVING THREATS IN TRANSPORTATION CYBERSECURITY

The rapid advancement of technology in CMV fleets and ADS has brought significant improvements in efficiency, safety, and operational control. However, these advancements also introduce new cybersecurity threats that are becoming increasingly sophisticated. Cybercriminals are continuously developing new attack methods, leveraging emerging technologies such as artificial intelligence (AI) to exploit vulnerabilities in connected transportation systems.

Future cybersecurity strategies must account for these evolving threats by incorporating advanced threat detection mechanisms, adaptive security frameworks, and stronger regulatory enforcement. Industry-wide collaboration will play a critical role in ensuring that cybersecurity defenses keep pace with the growing complexity of cyberattacks targeting transportation networks.

Increasing Sophistication of Cyberattacks on Transportation Networks

As transportation systems become more interconnected, the attack surface available to cybercriminals expands. Modern fleet operations depend on cloud-based fleet management systems, wireless communication networks, and real-time data processing, all of which can be targeted by cyber threats.

Recent trends in cyberattacks on transportation networks include:

- Advanced persistent threats where attackers infiltrate fleet management systems and remain undetected for extended periods, collecting sensitive data or preparing for large-scale disruptions.
- Ransomware attacks that target fleet operators, locking access to critical vehicle data or operational systems until a ransom is paid.
- Exploitation of zero-day vulnerabilities in fleet telematics software, allowing attackers to manipulate vehicle operations before security patches are released.
- Supply chain cyberattacks that introduce compromised software or hardware components into fleet systems, leading to hidden security vulnerabilities.

To mitigate these evolving threats, fleet operators must:

- Continuously update and patch software to address newly discovered vulnerabilities.
- Strengthen endpoint security on fleet devices to prevent unauthorized modifications.
- Enhance network segmentation to limit the spread of cyberattacks within fleet systems.
- Implement advanced IDS capable of recognizing complex attack patterns.

Emerging Threats from AI-Driven Cyberattacks

The growing use of AI in cybersecurity presents both opportunities and risks. While AI-driven security tools improve threat detection and automate response mechanisms, cybercriminals are also leveraging AI to conduct more sophisticated and adaptive attacks.

Emerging AI-driven threats include:

- AI-powered phishing attacks that generate highly personalized messages to deceive fleet personnel and extract login credentials.
- Automated malware that adapts its behavior to evade detection by traditional security systems.
- Deepfake technology used to impersonate fleet operators, leading to unauthorized access to fleet management platforms.
- AI-driven vulnerability discovery tools that allow attackers to rapidly identify and exploit weaknesses in fleet software.

To counter AI-driven cyberattacks, transportation cybersecurity strategies should:

- Deploy AI-based security analytics that can detect anomalous behavior in fleet networks and respond in real time.
- Use behavioral analysis tools to identify and block advanced phishing attacks.
- Implement deepfake detection mechanisms to verify the authenticity of communications and identity verification processes.
- Invest in adversarial AI research to anticipate and mitigate emerging attack methods before they become widespread.

Importance of Industry-Wide Collaboration and Regulatory Updates

Given the complexity and scale of modern transportation networks, cybersecurity cannot be addressed in isolation. Collaboration between fleet operators, government agencies, technology providers, and cybersecurity researchers is essential to developing effective security standards and best practices.

Key areas where industry collaboration and regulation play a role include:

- Establishing standardized cybersecurity frameworks for fleet operations, such as those outlined by the NHTSA and the CISA.

- Sharing threat intelligence among transportation industry stakeholders to stay ahead of emerging cyber threats.
- Developing secure-by-design principles for vehicle manufacturers, ensuring that cybersecurity is integrated into new vehicle architectures from the outset.
- Enforcing compliance with cybersecurity regulations, requiring fleet operators to implement minimum security standards for data protection and system integrity.

By fostering collaboration between industry leaders and regulatory bodies, transportation cybersecurity efforts can become more proactive, adaptive, and resilient against evolving cyber threats.

CONCLUSION

Cybersecurity in ground transportation is a growing concern as CMV fleets become increasingly dependent on ADS, networked telematics, and real-time data exchanges. The complexity of securing these systems requires a multi-layered approach that includes network security, data protection, physical security, continuous monitoring, and human factors considerations. Without robust cybersecurity measures, fleet operations are vulnerable to unauthorized access, system manipulation, and operational disruptions that could jeopardize safety and efficiency.

The integration of cybersecurity best practices in fleet management is essential to mitigating risks and ensuring the safe deployment of automated and connected vehicle technologies. As cyber threats continue to evolve, fleet operators must remain proactive in updating security protocols, implementing real-time monitoring solutions, and training personnel on cybersecurity awareness.

Summary of Key Cybersecurity Considerations for Ground Transportation

The cybersecurity challenges facing CMV fleets require a strategic combination of technical defenses, regulatory compliance, and industry collaboration. Key considerations include:

- Securing network communications and fleet management systems through encryption, firewalls, and IDS.
- Protecting data integrity by implementing strong authentication mechanisms and access control policies.
- Addressing human factor risks through ongoing cybersecurity training and awareness programs for drivers and fleet personnel.
- Enforcing physical security measures to prevent unauthorized access to vehicle hardware, telematics systems, and diagnostic tools.
- Establishing incident response plans and failure recovery strategies to minimize the impact of cyberattacks on fleet operations.
- Conducting continuous monitoring and penetration testing to identify vulnerabilities and strengthen fleet cybersecurity resilience.

By incorporating these cybersecurity strategies, fleet operators can reduce the risk of cyber incidents while ensuring that ADS-equipped vehicles and connected fleets operate safely and efficiently.

The Role of Continuous Research and Improvement in Cybersecurity for CMV Fleets

As cyber threats become more sophisticated, the transportation industry must prioritize ongoing research and development in cybersecurity. Threat actors are constantly finding new ways to exploit vulnerabilities in vehicle software, telematics systems, and cloud-based fleet management platforms. To stay ahead of these threats, fleet operators, researchers, and policymakers must continuously evaluate and enhance cybersecurity practices.

Key areas for ongoing research and improvement include:

- Advancing AI-driven threat detection systems that can identify and respond to cyberattacks in real time.
- Improving encryption and authentication protocols to protect fleet communications and prevent unauthorized access.
- Developing secure-by-design principles for ADS technology, ensuring that cybersecurity is integrated into vehicle systems from the initial development stages.
- Enhancing collaboration between government agencies, industry leaders, and cybersecurity professionals to establish standardized security frameworks for fleet operations.
- Investing in cybersecurity resilience testing, including red team exercises and simulated attack scenarios, to assess fleet security defenses and improve incident response capabilities.

By prioritizing continuous research and improvement, the transportation industry can build a more secure and adaptive cybersecurity infrastructure that safeguards CMV fleets against emerging threats.

Final Thoughts on the Balance between Automation, Security, and Human Oversight

The future of ground transportation relies on a careful balance between automation, security, and human oversight. While ADS technology and connected vehicle systems offer increased efficiency and safety benefits, they also introduce new cybersecurity challenges that require constant vigilance. Security measures must be designed to complement automation while ensuring that human operators remain actively involved in monitoring and responding to cybersecurity risks.

Automation alone cannot address all cybersecurity threats. Human oversight is critical in maintaining the security and integrity of fleet operations. Drivers, fleet managers, and cybersecurity professionals must work together to enforce security policies, recognize potential cyber threats, and implement rapid response strategies when needed.

Moving forward, achieving this balance will require:

- Strengthening cybersecurity education and training for fleet personnel to ensure they are equipped to handle cyber threats.
- Implementing fail-safe mechanisms that allow human intervention in ADS-equipped vehicles during cybersecurity incidents.
- Enhancing regulatory frameworks to address the evolving cybersecurity risks associated with automation and connected vehicle technologies.
- Encouraging industry-wide collaboration to share threat intelligence and develop standardized best practices for transportation cybersecurity.

Cybersecurity in ground transportation is an ongoing challenge that requires constant adaptation to emerging threats. By maintaining a proactive approach and integrating security into every aspect of fleet operations, the transportation industry can ensure a future where automated and connected vehicles operate safely, securely, and efficiently.

NOTES

1 Cybersecurity and Infrastructure Security Agency. (2019). *What is cybersecurity? Security Tip (ST04-001).* https://www.cisa.gov/uscert/ncas/tips/ST04-001#:~:text=Cybersecurity%20is%20the%20art%20of,integrity%2C%20and%20availability%20of%20information
2 Greenberg, A. (2015, July 21). Hackers remotely kill a Jeep on the highway—with me in it. *Wired.* https://www.wired.com/2015/07/hackers-remotely-kill-jeep-highway/
3 Tabak, N. (2020, December 28) *5 defining cyberattacks on trucking and logistics in 2020.* https://www.freightwaves.com/news/5-defining-cyberattacks-on-trucking-and-logistics-in-2020
4 Monteagudo, J. (n.d.). Cyber security for connected and autonomous vehicles. *Cyber Startup Observatory.*
5 National Institute of Standards and Technology. (2020). *Mobile device security: Corporate-owned personally enabled (COPE) (NIST Special Publication 1800-21).* U.S. Department of Commerce. https://doi.org/10.6028/NIST.SP.1800-21

Index

For Product Safety Concerns and Information please contact our
EU representative GPSR@taylorandfrancis.com Taylor & Francis
Verlag GmbH, Kaufingerstraße 24, 80331 München, Germany